P9-CJP-642

THE ATOM

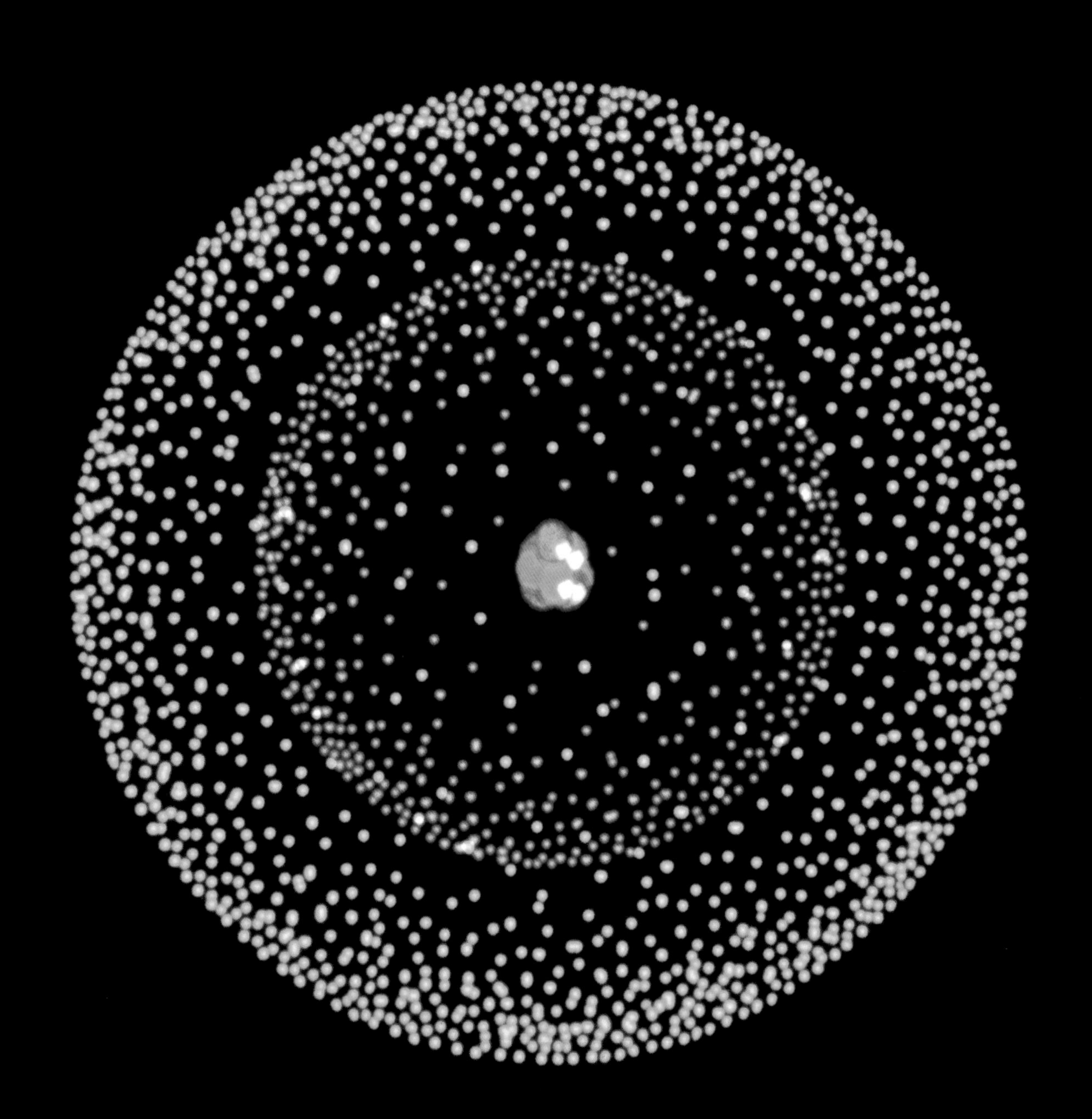

THE ATOM
A VISUAL TOUR

JACK CHALLONER

The MIT Press
Cambridge, Massachusetts

MIT Press edition, 2018

Note from the publisher: Although every effort has been made to ensure that the information presented in this book is correct, the authors and publisher cannot be held responsible for any injuries that may arise.

The MIT Press
Massachusetts Institute of Technology
Cambridge, Massachusetts 02142
http://mitpress.mit.edu

978-0-262-03736-5

Library of Congress Control Number: 2018933681

Printed in China

10 9 8 7 6 5 4 3

This book was conceived, designed and produced by
Ivy Press
An imprint of The Quarto Group
The Old Brewery
6 Blundell Street
London N7 9BH, UK

Publisher Susan Kelly
Creative Director Michael Whitehead
Editorial Director Tom Kitch
Commissioning Editor Kate Shanahan
Project Editor Stephanie Evans
Design Manager Anna Stevens
Designer Wayne Blades
Picture Researcher Katie Greenwood
Illustrator John Woodcock with additional illustrations by the author

Cover image: Computer graphic of an atom of beryllium
Kenneth Eward/Biografx/Science Photo Library

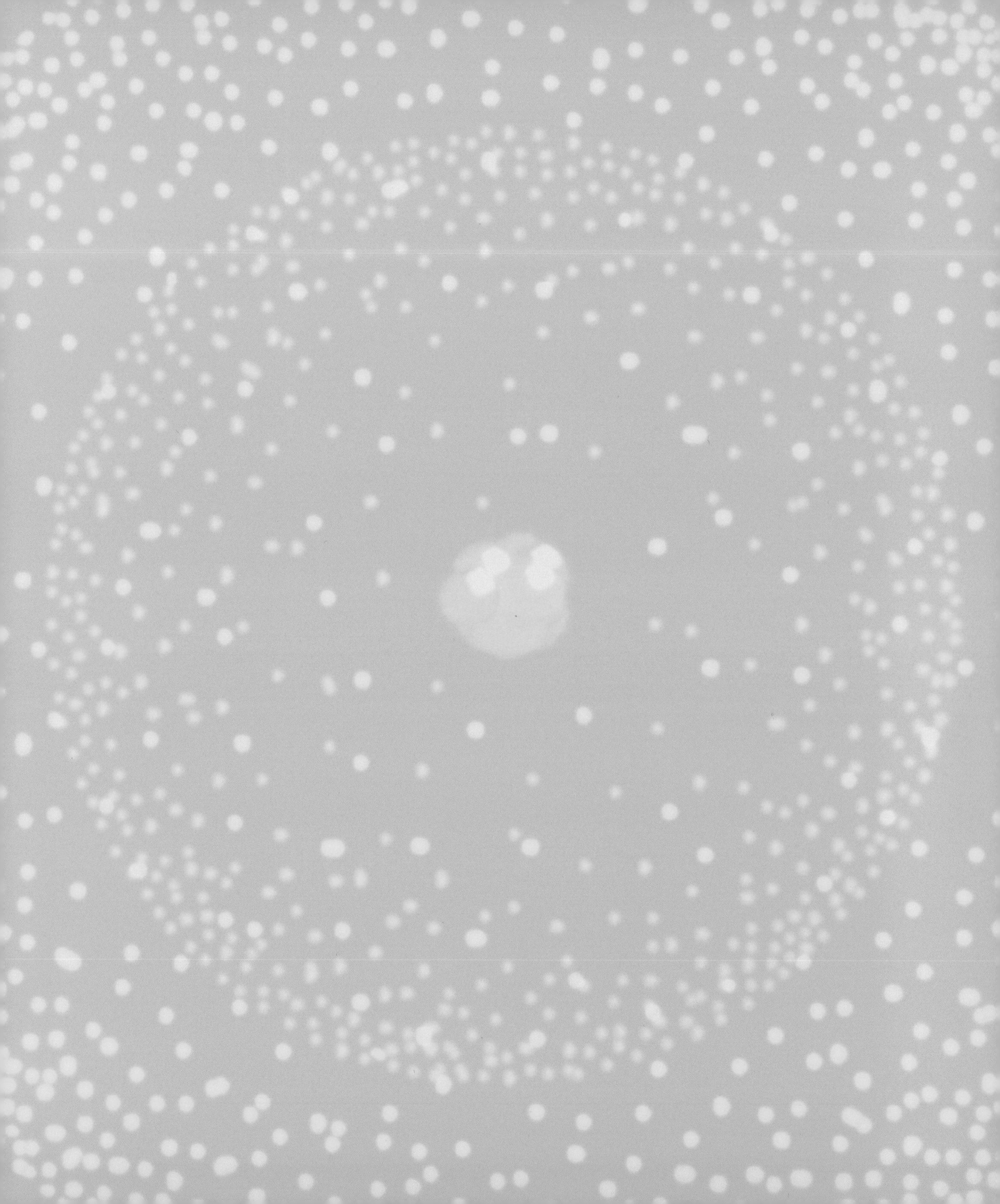

INTRODUCTION

From a distance, matter looks smooth and continuous. A solid object, such as a table, has definite edges and no apparent gaps in its structure; water spills from a cup as self-contained fluid streams and droplets; and the air we breathe, although invisible, feels like one gassy continuum. But at a scale far too small for us to truly comprehend, matter is bumpy and discontinuous; it is made up of empty space punctuated by countless tiny particles. This notion—that matter is made of extremely small particles—is called atomism. This book is a celebration of atomism, and it sets out some of the remarkable insights an atomic perspective has revealed about the world around us.

Despite the title of this book, and despite what the paragraph on the previous page might suggest, "the atom" is not actually the fundamental building block of matter. There are two reasons for this. First, an atom is itself made up of smaller particles: protons, neutrons, and electrons. Second, most matter is not actually made of atoms. The precise definition of an atom is of an isolated, self-contained object that has exactly the same number of protons as electrons—a situation that rarely exists in the real world. Strictly speaking, nearly all the substances around us are made of molecules or ions, not atoms. A molecule is made of two or more atoms joined together, but the atoms are entwined, sharing their electrons, not isolated and self-contained. An ion has unequal numbers of protons and electrons, so is not an atom either. Nevertheless, it is convenient to speak of matter as being "made of atoms." The atom is also a particularly useful archetype, a perfect starting point for understanding how matter works.

WHAT IS MATTER MADE OF?

People have probably wondered what matter is made of for as long as they have wondered about anything. The modern definition of the atom has developed as a result of theory, observation, and experiment over the past two hundred years; however, the idea that matter is made of extremely small particles is much older—though it has not always held sway. In the first chapter, we look back at the long history of the concept of "atoms," providing an overview of centuries of inspired work by brilliant minds.

The modern understanding of how atoms behave depends upon quantum theory, a counterintuitive but well-tested set of rules that govern the interactions of particles at the atomic scale. Chapter two explores quantum theory in some depth to make sense of the atom's basic structure. In this, its electrons are arranged in particular patterns around a dense central core, the nucleus, which is composed of the atom's protons and neutrons.

Only about ninety different kinds of atom exist naturally. Each has a different number of protons in the nucleus (always equal to the number of electrons arranged around it). Each kind of atom is called an element. A few elements were created in the first few minutes of the universe, as protons and neutrons bound together to form simple nuclei. Under the intense pressures and at the high temperatures inside stars, nuclei of these primordial atoms combine, or "fuse," to form larger nuclei, the basis of heavier elements. Other processes, in stars and in Supernovas—the extremely energetic explosions that befall large stars when they reach the end of their existence— create still heavier elements. Chapter three explains these processes that brought the elements into existence and explores the elements' properties. This chapter also introduces the periodic table, a way of organizing the elements into groups with similar properties.

Chapter four explores how the interactions between atoms can help to explain the physical and chemical properties of matter. The interactions of large numbers of particles (atoms, molecules, or ions) can explain the physical properties and behaviors of solids, liquids, and gases, such as air pressure, evaporation, and the existence of surface tension. Attractive forces between atoms create bonds: chemical bonds that produce chemical compounds. So, for example, atoms of hydrogen and oxygen bond together to form the chemical compound we call water.

In chapter five, we survey some of the modern technologies that enable researchers to gain intimate knowledge of atoms. These technologies produce stunning images of atomic surfaces or enable scientists to manipulate individual atoms. Chapter six explores some of the contributions that atomic theory has made to everyday life and beyond in the twentieth and twenty-first centuries. Quantum theory has, for example, given electronics engineers dominion over electrons, and that has led to the development of computers and other devices behind the digital revolution. Understanding nuclear reactions and radioactivity has opened up an enormous source of energy, which can be used for purposes both peaceful and otherwise.

the precise definition of an atom is of an isolated, self-contained object that has exactly the same number of protons as electrons

The final chapter of this book, chapter seven, reviews the current state of atomism, which is expressed in the Standard Model of particles and their interactions. This beautiful theory is the culmination of decades of work by theoretical and experimental physicists using powerful particle accelerators, such as the Large Hadron Collider at CERN on the Swiss–French border. The Standard Model makes sense of the huge menagerie of subatomic particles that exist and has made bold predictions that have been realized, such as the existence of the so-called God particle, the Higgs boson. At its heart are the true "atoms": the fundamental (un-splittable) particles from which the world is made. Within the Standard Model lies a conundrum, however, for it is based on quantum field theory—an extension of quantum theory in which particles are not solid objects, but instead manifestations of "fields" that permeate all of space. According to modern physics, then, "atoms" are nothing but waves drifting on an infinite, invisible, immaterial sea of potentiality.

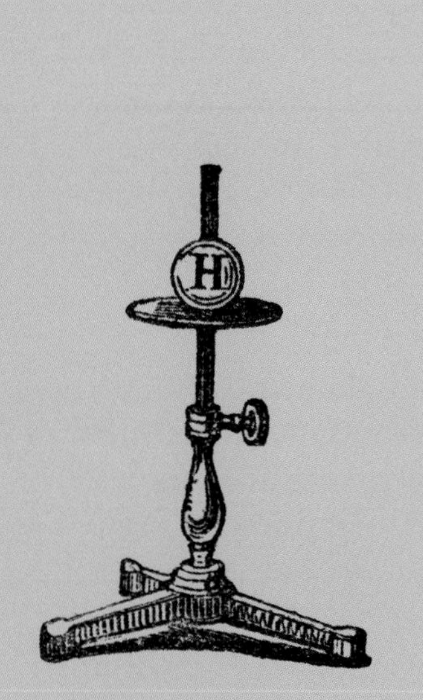
H

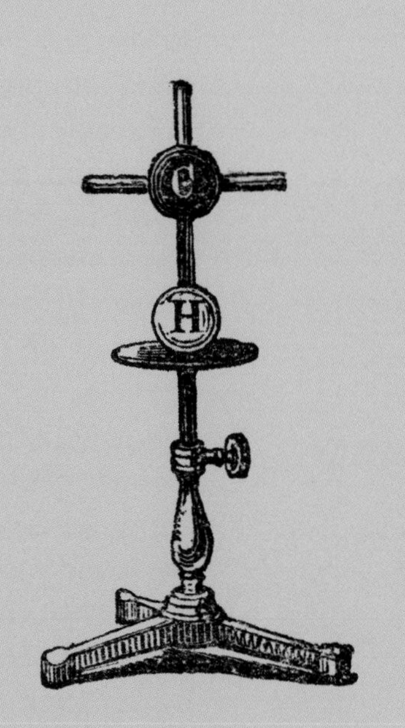
C
H

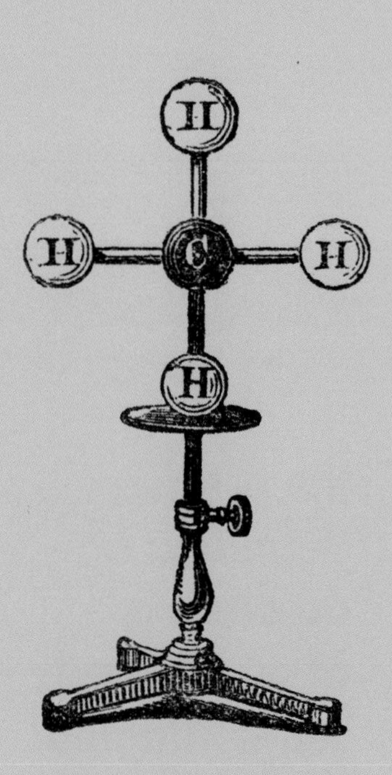
H
H
C
H
H

CHAPTER 1

A BRIEF HISTORY OF THE CONCEPT "ATOM"

The idea that matter is made of tiny particles is at least 2,500 years old. Through much of history, it was relegated to the fringes of scientific thought for philosophical and religious reasons. But it regained popularity with the rise of science in Europe in the seventeenth and eighteenth centuries. The notion that atoms really do exist only became widely accepted with the rapid rise of atomic physics in the early twentieth century.

Molecular models presented to the Royal Institution, London, UK, by German chemist August Wilhelm von Hofmann, in 1865. Hoffman used the models in a lecture entitled "The Combining Power of Atoms," at a time when the existence of atoms was still in doubt.

EVERYTHING AND NOTHING

The roots of modern atomic theory lie in ancient Greece. Oddly, perhaps, the story begins with philosophical wrangling about whether change is real or an illusion—and whether empty space can exist. Despite the development of well-thought-out and convincing atomic theories in both Greece and India, other ideas would come to dominate.

EVERYTHING CHANGES—OR DOES IT?

It was common for ancient Greek philosophers—as it is for scientists these days—to seek order in the world, and in particular to find a unified cause for the huge variety of phenomena we observe. When it comes to the physical world, or matter, the earliest Greek philosophers were "monists." They proposed that either all matter begins as one kind of substance and then differentiates, or that there actually is only one kind of matter, which manifests in various forms.

One of the earliest Greek philosophers, Thales of Miletus (ca. 625–ca. 545 BCE), suggested that water might be the primary substance from which all other substances derive. Anaximenes, also of Miletus, thought that the primary substance might be air.

Several decades later, another Greek philosopher, Heraclitus of Ephesus (ca. 535–ca. 475 BCE), suggested that the primary material might be fire. His reasoning was that fire is an agent of change—and, our senses report, change is a vital and constant feature of the world. A contemporary of Heraclitus, Parmenides of Elea (born ca. 515 BCE), believed quite the opposite. Parmenides and his followers rejected the empirical experiences of the senses, relying instead on pure reason. They believed that all change is an illusion, that it simply does not exist.

Parmenides's notion that change is an illusion followed from his belief that "nothingness" could not exist. He argued that the supposedly "changed" state of a thing is different from the original thing and so did not previously exist; it would therefore have to have come from nothing. Parmenides even rejected the idea that things could move. Motion was impossible, he said, because it would require the existence of "void," or empty space, into which an object could move—and void was the same as "nothingness." For Parmenides, reality was one perfect, full, unchanging sphere that had always existed and in which nothing ever changed. He called it the plenum. We will return to this notion in the context of modern theoretical physics in chapter seven.

ATOMS—MAKING SENSE OF CHANGE

The views of Parmenides were influential in ancient Greece, and the philosophers who succeeded him felt compelled to take his views into account. One of them was Democritus (ca. 460–370 BCE), widely credited with the first comprehensive atomic theory (although there were similar ideas in India at about the same time, see the box on page 14). Democritus attempted to reconcile Parmenides's notion of reality as an unchanging whole with the fact that change does seem to happen. He did so by making two adjustments to Parmenides's ideas. First, he suggested that "void," or empty space, can exist. Second, he proposed that all matter is made of tiny, indivisible particles. The individual particles retain their identity and their total number remain the same, so overall there is no change. But change can occur locally, because particles can move around, collide with one another, join and break apart, and rearrange themselves.

Democritus described his particles as ἄτομος (atomos, meaning "indivisible"). The word comes from ἀ- (a-, "not") and τέμνω (témnō, "I cut"). He also supposed that the particles are always in motion and that they are identical apart from their size and shape.

The doctrine of monism proposes that everything is made of, or arises from, just one kind of matter—but how can an ocean, rocks, soaring birds, and clouds be made of the same thing?

INDIAN ATOMISM

Around the same time that Democritus was formulating his atomic theory in Greece, Hindu, Buddhist, and Jain philosophers and religious thinkers in India were having very similar ideas. In early Jainism, for example, matter was considered to be one of six eternal substances, consisting of tiny, indivisible particles called paramāṇus. However, regardless of how well-formed the ideas of the Indian philosophers were and how similar they seem to Greek atomistic theories, it was the ideas of Greek philosophers that would influence the development of atomism in Europe, where many centuries later the modern atomic theory was born.

Democritus's atomistic philosophy attempted to make sense of the physical properties of matter. He suggested that a dense solid material is made of heavy and more closely-packed atoms, while a gas is made of extremely small, light atoms with a good deal of void between them. He also proposed that atoms link together, with physical connections, such as hooks and eyes, and that those connections can break and reform when chemical reactions take place, or when liquids evaporate and vapors condense—themes we explore in chapter four. Furthermore, the shapes of atoms confer certain properties; atoms of a liquid, for example, are round, so they easily flow over one another, while the atoms of salt are sharp.

Democritus's theory was not widely accepted, for two main reasons. First, it is a purely materialistic vision of the world; there is no room for metaphysical or spiritual influences. Democritus had envisaged a special type of atom for the soul; these atoms are smaller than the others, able to pass easily between the atoms of the body. The materialistic character of his theory made it unpopular with many people, especially religious thinkers—for how can one reduce the human spirit and imagination to the movement of atoms?

NOTHING REALLY MATTERS

The other main sticking point for Democritus's theory was its reliance on the notion of void—empty space. This would become increasingly important because of the ideas of one man: Aristotle (384–322 BCE). Aristotle's ideas about matter were pragmatic, based very much on experience of the world, which is one reason why his works were so influential. Aristotle believed that matter is continuous and, in principle, infinitely divisible. The character, or "form," of a substance is a separate quality from the matter itself. He stated his ideas as if they were truths, and for centuries that is how most scholars accepted them.

Aristotle firmly believed that empty space cannot exist. Any empty space would immediately be filled by matter around it, he claimed. Perhaps his most famous phrase is *horror vacui*, normally translated as "nature abhors a vacuum." Because the existence of void is such a crucial element of Democritus's theory, a belief that void cannot exist amounted to a strong denial of atomism.

Democritus (left) was born in Abdera, in Thrace, in the fifth century BCE. He wrote on a huge range of subjects, including mathematics, ethics, aesthetics, and epistemology (theory of knowledge), although almost none of his works survives.

Aristotle (above), a scholar of the fourth century BCE, was one of the most prolific and influential of the Greek philosophers. He wrote on a wealth of topics, including physics, biology, astronomy, weather, ethics, politics, poetry, theater, and language.

A SCHOLASTIC VIEW

Philosophers and other scholars across the early Islamic world and in medieval Europe largely accepted Aristotle's views on matter without question. Because Aristotle did not believe that matter is made of tiny particles, atomism remained largely in the shadows. But a new spirit of scientific investigation in Europe in the seventeenth century gave a new lease of life to the idea.

Knowledge and ideas from ancient Greece and ancient India spread widely from the third century BCE onward, as empires rose and fell—including during the Hellenistic period formed in the wake of conquests by Alexander the Great (356–323 BCE), and, of course, the Roman Empire. Aristotle's philosophy was championed by early Christian scholars, and also by Arabic scholars from the eighth century onward, in a flourishing empire that was

united by a common language (Arabic) and a common religion (Islam). The Islamic empire, characterized by a succession of powerful caliphates and dynasties, extended across a vast region, including the Arabian Peninsula, parts of India, the Middle East, North Africa, southern Italy, and Spain.

THE ISLAMIC GOLDEN AGE

The period from the ninth to the twelfth centuries is often called the Islamic Golden Age, because of the high culture and sophisticated state of science, mathematics, and engineering across the Islamic empire at that time. Many Arabic scholars absorbed, translated, and reinterpreted the classical works from ancient Greece, as well as making their own contributions in a wide range of subjects. Several scholars developed their own atomic theories, most notably Abū al-Ghazālī (ca. 1058–1111). He believed that all matter was made of indivisible particles that were arranged by Allah.

Two scholars in particular were responsible for saving and promoting the works of Aristotle. They were Ibn Sīnā (known in Europe as Avicenna; ca. 980–1037) and Ibn Rushd (known as Averroes; 1126–98). Both men rejected the kind of atomism proposed by Democritus, and both were highly influential for centuries to come.

INTO EUROPE

Knowledge amassed by Arabic scholars passed to Europe, mainly via Spain, and was adopted into the "scholastic" system—the teaching method used in the new universities of Europe from the eleventh century onward. As far as theories of matter are concerned, it was Aristotle's ideas that dominated. Aristotle considered matter to have substance and form. The form can change, but the substance remains the same. Most important, matter is continuous and void cannot exist.

Despite the prominence of Aristotelian thinking, the concept of atomism lived on for two main reasons, both of which derived from Aristotle's own work. First, Aristotle discussed Democritus's theory extensively in his writing, albeit critically. In fact, Aristotle is one of our most important sources of knowledge about Democritus, because none of his works has survived. Second, Aristotle did discuss the "smallest parts" of matter, which he called *minima naturalia*. Aristotle's "minima" were not indivisible particles like the atoms of Democritus; instead, they were the smallest amount of a particular substance. Aristotle wrote, for example, that you can divide flesh into increasingly smaller pieces. Below a certain size, however, the matter would still exist but it would no longer be flesh.

Aristotle did not dwell or particularly elaborate on the concept of minima, and so left room for interpretation. Some medieval scholastics adapted the idea with a leaning toward atomism. The influential Italian scholar Julius Caesar Scaliger (1484–1558) was generally a fierce defender of Aristotle's ideas at a time when some were beginning to question them, but he considered Aristotle's minima to be physical objects, the building blocks of matter, not just a limit to the division of a substance. He wrote, for example, about how water can wear away stone, one particle at a time, and how substances are more or less dense because their minima are closer together. There were even some scholars who fully supported atomism in the spirit of Democritus, but they were few. The rise of atomism proper had to wait until scholars began to question Aristotle's ideas fully to come up with their own ideas and conduct experiments to test them.

Scholasticism was an approach to learning philosophy and theology in European universities, based on the works of Aristotle and the teachings of early Christian thinkers. It was purely didactic, a way of passing on learning, without much room to assess critically what was being learned.

QUESTIONING ARISTOTLE

During the Renaissance, which began in the fourteenth century, artists, writers, philosophers, mathematicians, and scientists beyond the scholastic system gained a new interest in the culture of ancient Greece and Rome. These bold thinkers began challenging dogma and creating their own culture in the spirit of the Greeks and the Romans. Johannes Gutenberg's printing press, invented about 1440, helped spread these new ideas far and wide. Among many important advances, the Renaissance spawned the works of Leonardo da Vinci and Michelangelo, Nicolaus Copernicus's theory about Earth orbiting the Sun, Andreas Vesalius's remarkable work on human anatomy, and the Protestant Reformation.

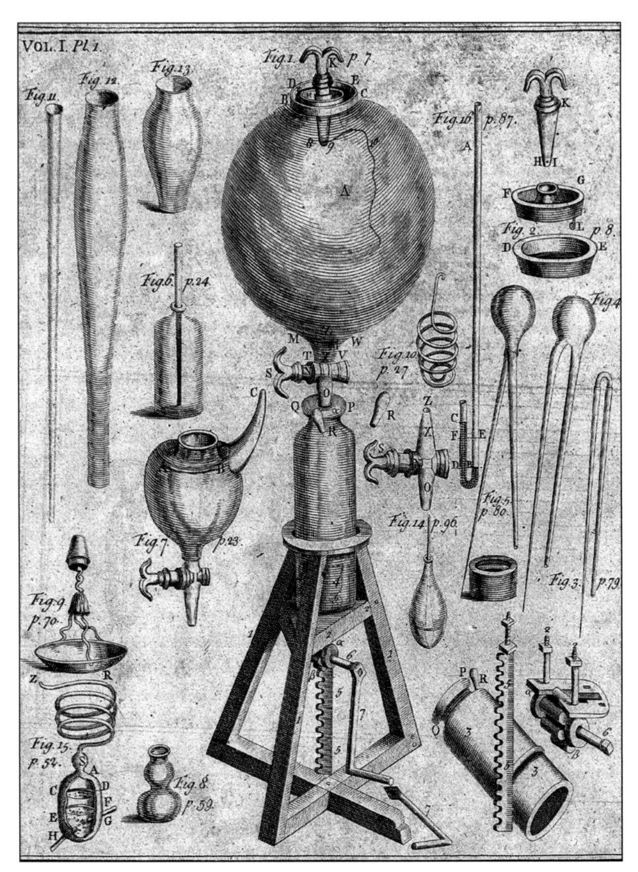

By the early seventeenth century, ideas about how the world works and what it's made of had shifted from the philosophical to the scientific, with an emphasis on empiricism—a reliance on experience, on an investigation of the real world reported by our senses. In the hands of people such as Galileo Galilei (1564–1642), the newly invented telescope and microscope were highlighting problems with Aristotle's views, fueling the new spirit of investigation. And in 1620, English politician and scientist Francis Bacon (1561–1626) codified the scientific method, in his book *Novum Organum*; the title in reference to Aristotle's *Organon*, a six-volume work on how to obtain knowledge using logic. With scientists observing, questioning, theorizing, and experimenting across Europe, it was no wonder that views on what matter is made of began to change.

By the mid-seventeenth century, there were several prominent scientists who believed matter is made of particles. One new phenomenon that helped encourage the rise of atomism was the vacuum, the void so vehemently dismissed by Aristotle. In 1643, Italian scientist Evangelista Torricelli (1608–47) observed what seemed to be empty space inside the glass of his mercury barometer. In 1654, German scientist and politician Otto von Guericke (1602–86) invented a crude vacuum pump, able to create partial vacuums inside sealed containers. During the 1660s, von Guericke carried out hundreds of experiments on the partial vacuums he created. His work inspired Anglo-Irish scientist Robert Boyle (1627–91) to develop a much more powerful and effective pump that could produce a much better vacuum. Boyle would also be a pioneer in another important strand in the history of atomism: the science of chemistry.

Robert Boyle's vacuum pump, along with some of the apparatus he used to experiment with low pressure air.

Two hemispheres held together by atmospheric pressure, after von Guericke removed the air from them, could not be pulled apart by teams of horses in a demonstration at Regensberg, Germany, in 1654.

Torricelli's barometer was a long glass tube sealed at one end and filled with mercury, a dense metal that is liquid at room temperature. When the tube was held upright with the open end submerged in a bowl of mercury, the mercury inside the tube made a column about 30 inches (76 centimeters) tall. Above that was apparently empty space: a vacuum.

A NEW SCIENCE OF MATTER

Founded late in the seventeenth century, the science of chemistry gradually took over its nonscientific predecessor, alchemy. Chemists founded scientific laws about the behavior of matter that would make the rise of atomism inevitable. Inspired by new knowledge about elements and compounds, English chemist John Dalton founded the first modern atomic theory at the beginning of the nineteenth century.

CHEMICAL REACTIONS

Combustion produces fire and reduces wood to a pile of ashes; fermentation changes plant matter into wine or beer; and smelting produces metals from rocks. Without knowledge of atoms and an understanding of elements and compounds, chemical reactions such as these are mysterious—even magical. Before the science of chemistry began to unlock what is really going on during chemical reactions, philosophers and other scholars based their understanding on the ancient art of alchemy.

Several distinct types of alchemy evolved in several different parts of the world. European alchemists adopted and developed their practices and beliefs from Arabic alchemists of the Islamic Golden Age. The Arabic alchemists developed many of the fundamental laboratory techniques and much of the basic equipment found in chemical laboratories today.

Arabic and European alchemists saw chemical changes as transmutations, substances "changing form." Their thinking was based on Aristotle's idea that a substance is made of matter plus form (see page 17). They also based their thinking on the ancient theory of the four "elements"— earth, air, fire, and water—a theory that Aristotle had also championed and developed. These four substances are not chemical elements as defined by modern science. Some alchemists added mercury and sulfur to the list of elements. Ironically, perhaps, these really are chemical elements. Alchemists saw transmutation as a shift in the proportions of elements within a substance.

According to the theory of the four elements followed by the alchemists, wood is made mostly of fire, water, and earth. When wood burns, the "fire" and "water" components are released, leaving behind earth, which has the appearance of ashes. The modern understanding of the chemical reactions taking place is based on the breaking and making of bonds between atoms in the wood and the water.

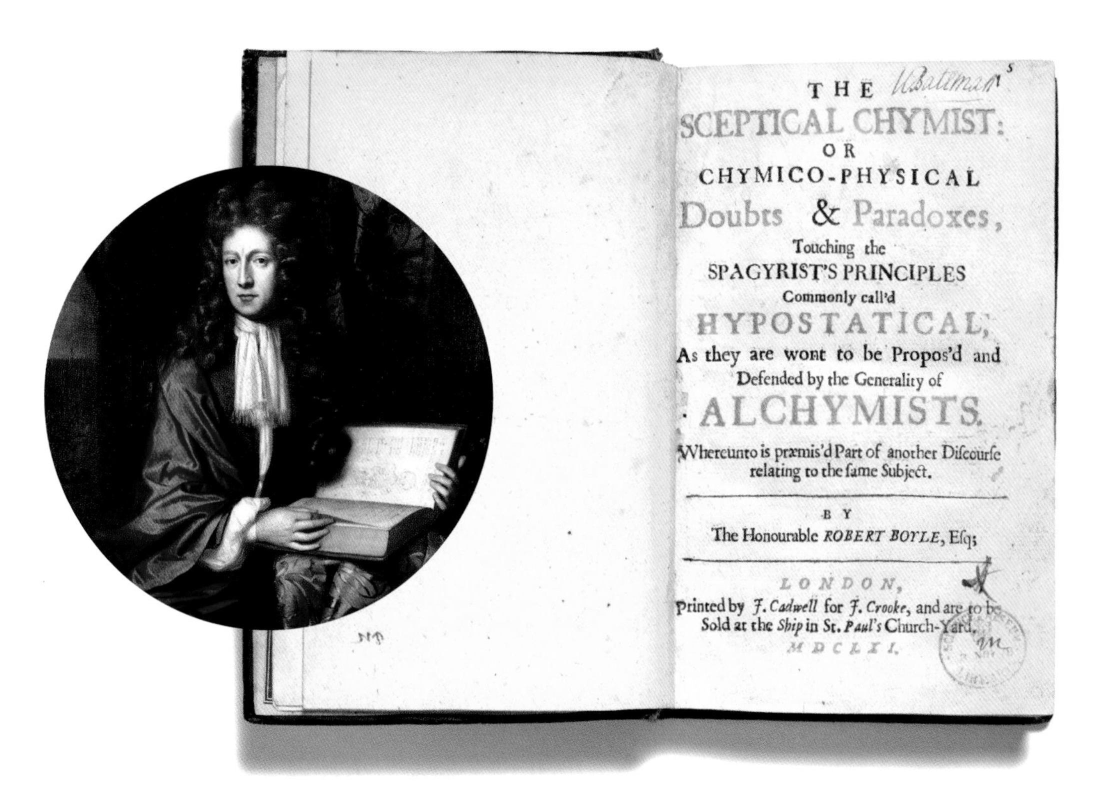

THE
SCEPTICAL CHYMIST:
OR
CHYMICO-PHYSICAL
Doubts & Paradoxes,
Touching the
SPAGYRIST'S PRINCIPLES
Commonly call'd
HYPOSTATICAL;
As they are wont to be Propos'd and
Defended by the Generality of
ALCHYMISTS.
Whereunto is præmis'd Part of another Discourse
relating to the same Subject.

BY
The Honourable ROBERT BOYLE, Esq;

LONDON,
Printed by J. Cadwell for J. Crooke, and are to be
Sold at the Ship in St. Paul's Church-Yard.
MDCLXI.

In *The Sceptical Chymist,* Robert Boyle (inset) encouraged people to take a more "philosophical" approach to investigating matter. At the time, most people who used chemical reactions were simply following recipes to make pharmaceuticals.

ELEMENTS OF CHANGE

Partly as a result of his experiments with vacuums, but more as a result of his firm belief in empiricism, Robert Boyle came to the conclusion that matter must be made of particles. He made his views clear in *The Sceptical Chymist* (1661), an influential book that helped to found the new science of chemistry.

Boyle adhered to a form of atomism that was gaining popularity in the seventeenth century: corpuscularianism (see page 22). In the context of this belief, Boyle challenged the theory of the four elements and proposed a new definition for an element:

"certain Primitive and Simple, or perfectly unmingled bodies; which not being made of any other bodies, or of one another, are the Ingredients of all those call'd perfectly mixt Bodies are immediately compounded, and into which they are ultimately resolved."

In other words, an element is a substance made of just one kind of particle, and the particles of elements mix together to form other, compound, substances. Boyle suggests that it is possible, in principle, to obtain pure elements by breaking down everyday compounds into their components. More important than their support for atomism, and also more important than the new definition of an element, Robert Boyle's books heralded a new, scientific approach to investigating chemical reactions and matter in general.

CORPUSCULARIANISM

To the corpuscularians, matter was undoubtedly made of particles. However, there was no definite opinion on whether the particles fill space (like the plenum of Parmenides, see page 12) or are surrounded by void (suggested by Democritus, page 13). In the seventeenth and eighteenth centuries, there were several influential corpuscularians, among them French philosopher and mathematician René Descartes (1596–1650) and English scientist Isaac Newton (1642–1727). In his book *Opticks* (1704), Newton set out at length how the corpuscles of matter might explain physical and chemical reactions. Having discovered that gravity seemed to act as a "force at a distance," he wondered whether the particles of matter might be held together by a similar kind of force. He also suggested that when particles of a liquid evaporate to become a gas, they remain in contact with each other but grow to many times their original size and become springy, explaining why gases can be compressed.

Also in *Opticks*, Newton suggested that even light is made of corpuscles. Dutch scientist Christiaan Huygens (1629–95) reached a different conclusion about light: that it travels as waves. But in his book *Traité de la Lumière* (Treatise on Light, 1690), he clearly showed that he was a corpuscularian when it came to matter, explaining the regular shapes of crystals (shown below) as being a result of their being composed of corpuscles of matter:

It seems in general that the regularity that occurs in these productions comes from the arrangement of the small invisible equal particles of which they are composed.

However, he admitted that many mysteries remained:

I will not undertake to say anything about how so many corpuscles, all equal and similar, are generated, nor how they are set in such beautiful order.

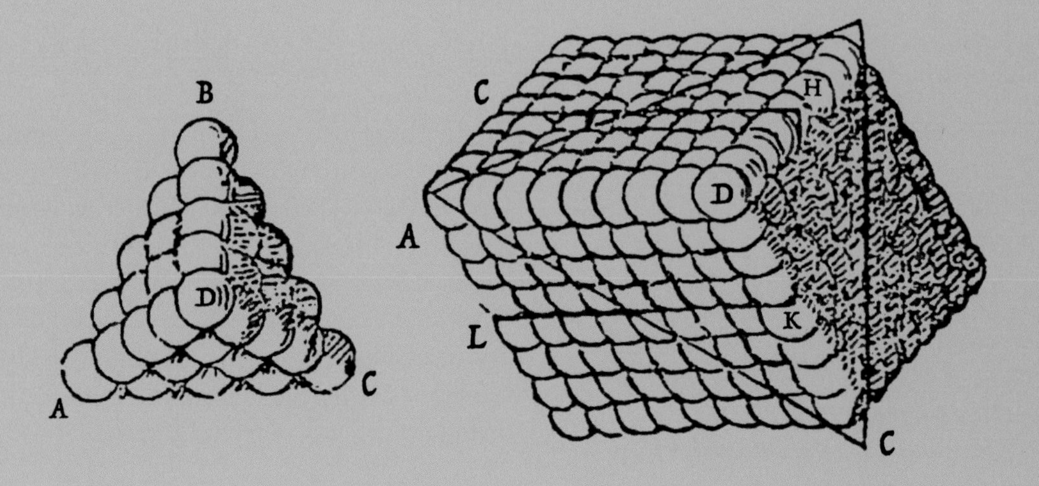

CHANGE IN THE AIR

One area of chemistry that received particular attention from empirical science was the study of gases, or "airs." In 1640, Flemish scientist Jan van Helmont (1580–1644) coined the term "gas," based on the Greek word *khaos* (meaning "empty space"), as the name of an "air" he discovered—a gas we now know as carbon dioxide.

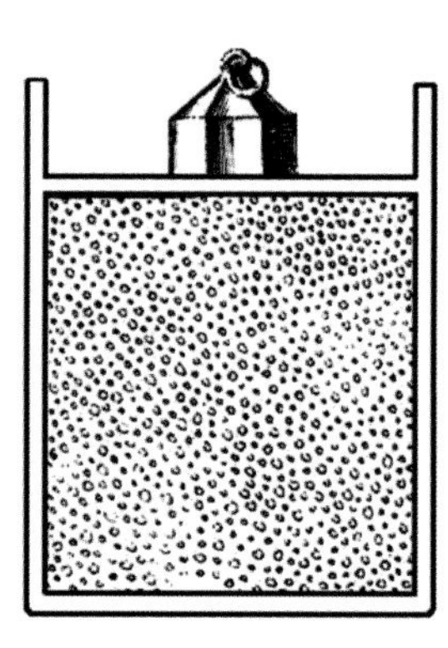

A drawing from Daniel Bernoulli's *Hydrodynamica* illustrates his idea that gases are made of tiny particles in motion.

In the second half of the seventeenth century and the first half of the eighteenth, scientists studied mostly the physical rather than chemical properties of gases. In particular, they worked out the relationships between the pressure, temperature, and volume of gases. In 1662, Robert Boyle discovered that if the temperature of a gas is held constant, its pressure and temperature are inversely proportional to each other. In other words, double the pressure on a gas (compress it twice as hard) and you will halve its volume, and vice versa. This relationship became known as Boyle's Law. Boyle pictured a gas as made of stationary corpuscles with springs between them. Other scientists worked out similar laws that related pressure, temperature, and volume.

Swiss mathematician Daniel Bernoulli (1700–82) pictured gases differently. In one chapter of his 1738 book *Hydrodynamica*, Bernoulli suggested that the particles of a gas are tiny solid balls, not large, springy corpuscles. The tiny balls traveled at high speed and collided with each other and with the walls of the container that confined them. The many impacts the particles made on the container walls would account for the pressure exerted by the gas. Bernoulli even related the temperature of the gas to the average speeds of the balls. His mathematical approach led him to an equation that was in excellent agreement with Boyle's Law and the other gas laws.

Despite this brilliant insight, few of Bernoulli's contemporaries took any notice of what should have been a crucial piece of evidence in favor of atomism—largely because they could not see why the tiny solid balls would not gradually run out of energy and all end up at the bottom of the container. However, chemists continued to work on gases with great fervor. In the 1750s, Scottish chemist Joseph Black (1728–99) found new ways to produce carbon dioxide, which he named "fixed air." Research into gases led to the discoveries of "flammable air" (hydrogen, 1766), "phlogisticated air" (nitrogen, 1772). and "dephlogisticated air" (oxygen, 1773).

French chemist Antoine Lavoisier (1743–94) took great interest in these gases, and in particular in the chemical reactions that produced them and in which they took part. Lavoisier was an extremely meticulous experimenter who carefully weighed the chemicals before and after a reaction, including the gases. His research led him to several important conclusions that would firmly establish chemistry as a science, and it would also prepare the ground for the modern atomic theory. Lavoisier found that the total mass of the chemicals taking part in a chemical reaction was exactly the same as the total mass of the reaction products—a phenomenon known as the conservation of mass. He also worked out that combustion involved dephlogisticated air (oxygen) combining with other substances. When flammable air (hydrogen) burned, it combined with dephlogisticated

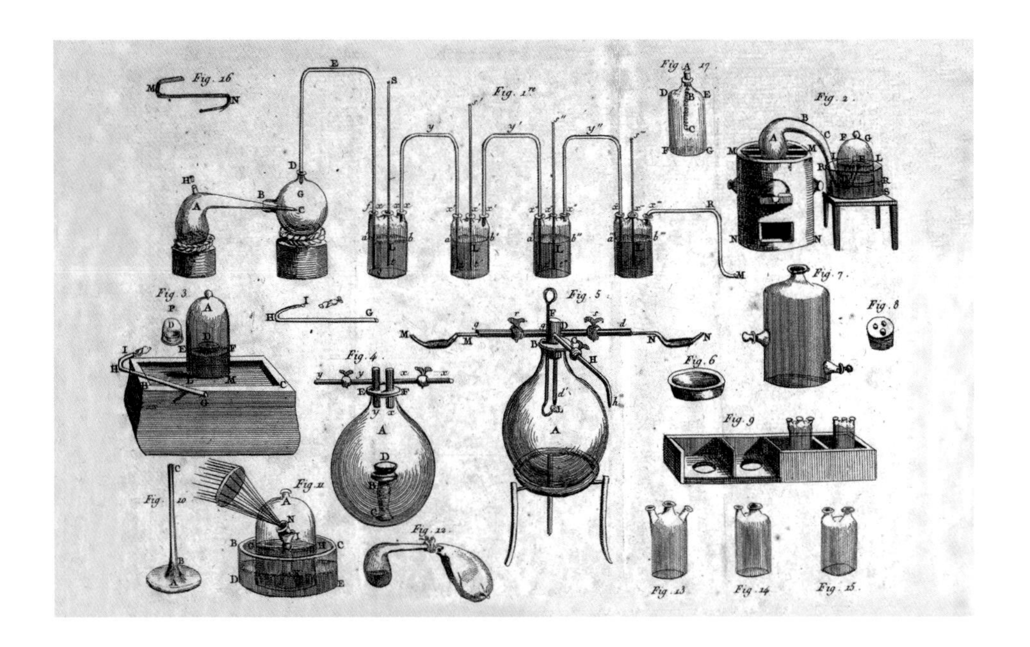

A selection of Antoine Lavoisier's laboratory equipment, as shown in his book *Traité Élémentaire de Chimie* (1789).

air to produce water. Lavoisier himself came up with the name "hydrogène," meaning "water maker." He had found empirical evidence to support Boyle's idea; hydrogen and oxygen are elements that join together to form a compound, water. He even produced the first list of chemical elements, which he published in his book *Traité Élémentaire de Chimie* (Elementary Treatise on Chemistry, 1789). In addition to hydrogen and oxygen, Lavoisier's list also included nitrogen, phosphorus, sulfur, and seventeen metallic elements. But he also included some substances that are actually compounds, not elements—and his list even included light and heat.

In the nineteenth century, in the spirit of Boyle and Lavoisier, chemists began to discover many more previously unknown elements, giving more credence to Boyle's definition, and more weight of argument against Aristotle's ideas.

A NEW ATOMIC THEORY

Lavoisier's insight into the relationship between elements and compounds, and his penchant for accurate chemical analysis, led French chemist Joseph Proust (1754–1826) to investigate an idea that most chemists of the time assumed to be true, but which had never been tested. The idea was that when elements form compounds, they do so in definite proportions. So, for example, if 100 grams of a compound contains 60 grams of one element and 40 of another, then in a 200-gram sample of the same compound there would be 120 grams of the first element and 80 grams of the second. (Note how the total mass of elements equals the mass of the compound they produce, in accordance with Lavoisier's conservation of mass.) In the 1790s and 1800s, Proust carried out a large number of experiments on many different compounds, establishing the truth of what chemists had suspected, in his "Law of Definite Proportions."

Meanwhile, English chemist John Dalton (1766–1844) was studying gases and, in particular, how mixtures of gases behave. He was already convinced that matter is made of particles, and, in the first few years of the 1800s, he was struck by a simple yet powerful qualification of that idea. He realized that every atom of a particular element must have exactly the same mass as all the others, and that this "atomic weight" would be different for each element. The ratio of the masses of elements in a compound are simply the ratios of the atomic weights of the elements involved. Dalton envisaged compounds as being made of molecules, each made of two or more atoms joined together. This vision made sense of Proust's Law of Definite Proportions; if it holds true for a single molecule, it would hold true for any number of molecules, even in a sample large and heavy enough to weigh.

Dalton went farther. He realized that in some cases, there might be two or three distinct ratios, corresponding to distinct compounds made up of different combinations of the same elements. So, for example, mercury combines with sulfur to form two compounds, which we now call mercury (II) sulfide (HgS), and mercury (I) sulfide (Hg_2S) —and the ratios of mercury to sulfur by mass are 25:4 and 50:4. This was Dalton's Law of Multiple Proportions, and there seems no reasonable explanation of it other than the notion that compounds are made of atoms with definite weights grouped together.

Dalton discussed and outlined the first robust, modern atomic theory in *A New System of Chemical Philosophy* (1808). He proposed that atoms are indivisible, and cannot be created or destroyed; all atoms of an element are of the same size and mass, and they have the same properties; atoms of different elements combine in simple ratios to form chemical compounds; and chemical reactions involve a rearrangement of atoms. In his book, Dalton listed the known elements with their estimated atomic weights, relative to the lightest of the elements, hydrogen—the first time anyone had done this. Dalton's theory united the physical and chemical aspects of matter and laid a firm foundation for modern atomic theory. But it did not become mainstream science for many decades.

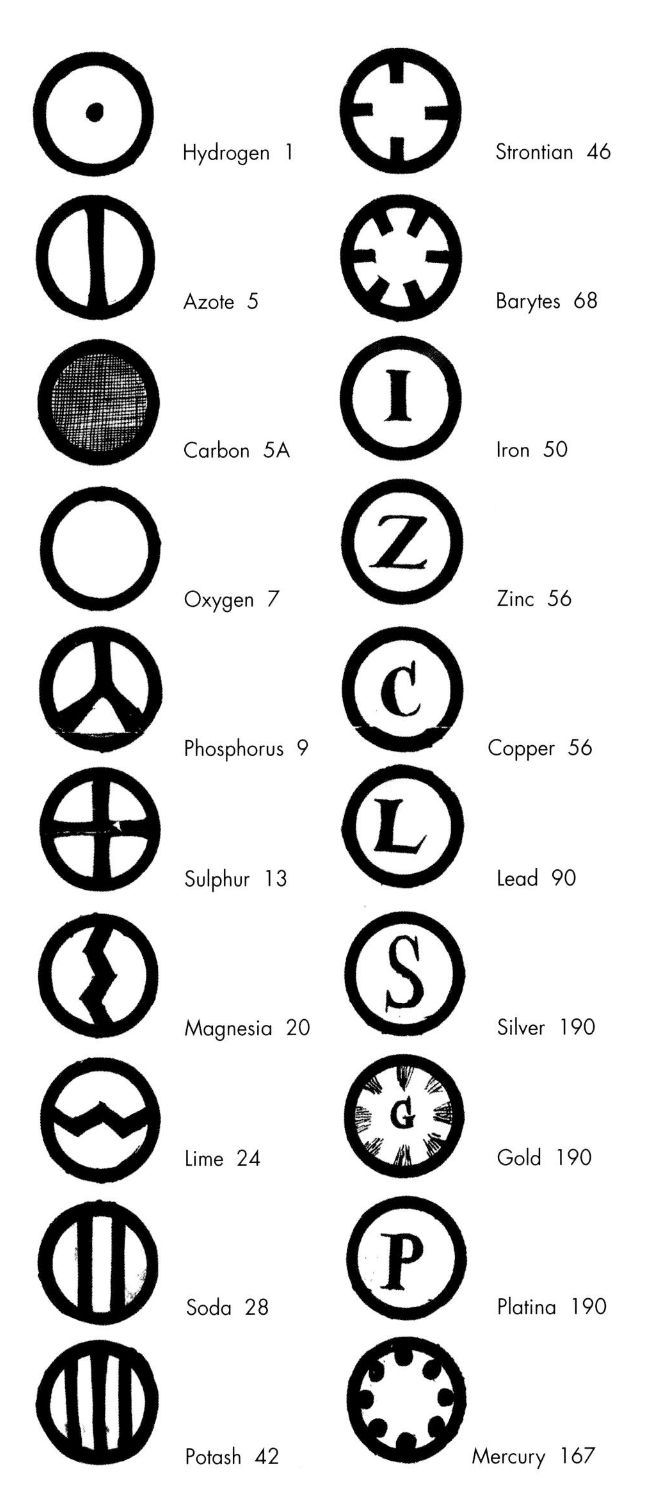

John Dalton used circular symbols for the elements, emphasizing his atomic theory. For Dalton, all the atoms of a particular element had the same mass (the numbers alongside the names, relative to hydrogen, 1). Note that some of the "elements" he lists are actually compounds.

A CONVINCING THEORY

Despite John Dalton's compelling theory, many nineteenth-century scientists did not believe that atoms could actually be real. But mounting evidence in favor of atomism, both empirical and theoretical, convinced more and more chemists and physicists. By the beginning of the twentieth century, nearly all scientists had accepted that atoms really do exist.

CHEMISTRY—BUSINESS AS USUAL

John Dalton's atomic theory provided a robust framework for understanding elements, compounds, and chemical reactions in terms of tiny particles. But chemists had little need of a physical theory of matter, so the atom remained a hypothetical object in the eyes of many of them. Despite this, they did measure the "atomic weights" of elements they discovered, and atoms were routinely discussed.

Mendeleev's handwritten periodic system of the elements (below) is dated February 17, 1869. In honor of Mendeleev (pictured above right), in 1955 a newly discovered element was named mendelevium (Md).

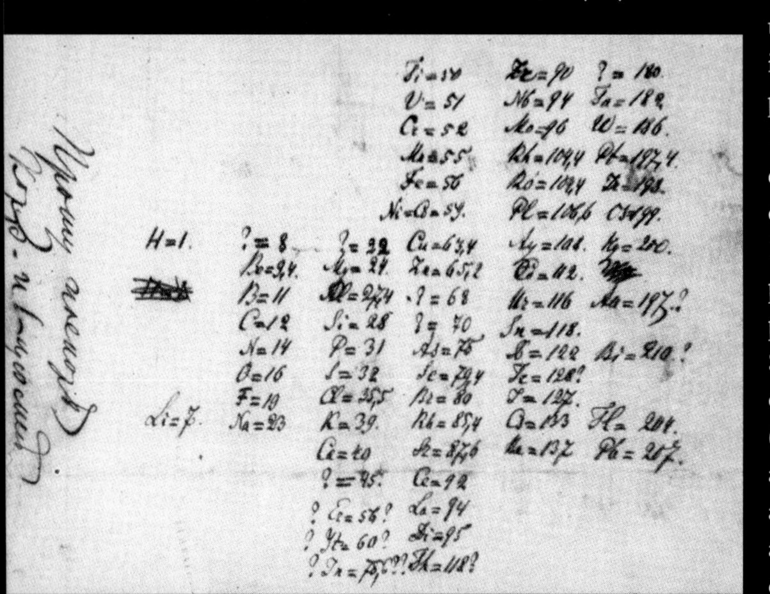

Meanwhile, chemists continued to discover new elements and to investigate a huge range of chemical reactions. One important new tool available to chemists was the electric battery. Investigations of matter using electricity hinted that matter itself has electrical properties. Several scientists used batteries to separate water into hydrogen and oxygen in 1800, just a few months after the invention of the device. English chemist Humphry Davy (1778–1829) used a powerful battery to decompose many compounds into their elements, discovering several elements in the process, among them sodium and potassium (both in 1807). Davy's colleague Michael Faraday (1791–1867) carried out extensive research into the effect of electricity on chemical compounds dissolved in water or acids. In 1834, he coined the term "ion" for an electrically charged particle that would migrate toward one or other electrodes placed in a solution. Decades later, Swedish chemist Svante Arrhenius (1859–1927) used the term in his theory of how compounds dissolve in water: by dissociating (breaking down) into ions. Arrhenius suggested that atoms contain several charged objects,
and that these could be lost or gained, giving the atom a net positive or negative charge. He had predicted the existence of the electron.

Another important tool for chemists was the spectroscope: a device that allowed for careful analysis of the spectrum of light emitted by objects. In the 1820s and 30s, several scientists noticed bright lines in the spectrum of the light given off by certain chemical compounds heated in flames. The particular pattern of lines was characteristic of the elements that were present in the compounds. German chemist Robert Bunsen (1811–99) and physicist Gustav Kirchoff (1824–87) systematically recorded the spectrum of each of the known elements. They discovered the elements cesium (1860) and rubidium (1861) when they observed patterns of bright lines that had not been seen previously. In the hands of other scientists, Bunsen and Kirchoff's approach was to lead to the discovery of several other elements. Among them was helium, discovered by studying the spectrum of the Sun in 1868, decades before it was found on Earth.

By the 1860s, more than fifty chemical elements were known, and scientists began to notice patterns in their behavior. The elements seemed to fall into groups according to their physical properties, as well as their chemical properties, such as the type of compounds they formed. English chemist John Newlands (1837–98) and German chemist Julius Meyer (1826–1909) noticed that the members of a particular group seemed to be at regular intervals, or periods, in the list of elements when arranged by increasing atomic weight. Meyer constructed a table of some of the elements, with similar elements grouped together. But the periodicity was not perfect, not quite regular enough. Russian chemist Dmitri Mendeleev (1834–1907) was so confident in the periodic nature of the list of elements that he simply left gaps in his "periodic table" to represent elements as yet undiscovered, so that these elements could fall into the right groups. Over the next few decades, Mendeleev was proved right; all the elements he had predicted were discovered. We will examine the periodic table, and how it relates to the inner workings of the atom, in chapter three.

Compounds containing metal held on a wire in a hot flame emit light with colors characteristic of the metallic elements present. Above are: strontium (1), copper (2), and potassium (3). Splitting the light with a spectroscope reveals bright lines at certain parts of the spectrum. This emission spectroscopy was used in the discovery of several previously unknown elements in the nineteenth century.

PHYSICISTS ON ATOMS

While chemists were busy discovering and cataloging new elements, nineteenth-century physicists were trying to make sense of basic phenomena, such as light, heat, electricity, and magnetism. In the previous century, scientists had seen each of these phenomena as a distinct "imponderable fluid" that surrounds or penetrates objects and could be transferred between them. Heat fluid, for example, would flow from a flame to a cooking pot. Increasingly, however, new theories and evidence were undermining the imponderable fluids idea. For example, the discovery of electromagnetism in 1820 showed that electricity and magnetism are intimately linked, so they cannot be separate "fluids."

Electromagnetism is also intimately related to light. In the 1860s, British physicist James Clerk Maxwell (1831–79) found that light is a wave of undulations in electric and magnetic fields. He had discovered that light is a form of "electromagnetic radiation," which is produced whenever electric charges are accelerated (their speed or their direction of movement is changed). Light, then, was also not an imponderable fluid.

The idea of heat as a fluid had also been overturned. In its place physicists revisited a notion suggested by several scientists in the past—that heat and temperature are related to the movement of the particles of matter. Heat up a solid substance, and you increase the vibration of the particles of which it is made. If you heat it sufficiently, the particles can break away from each other just enough to move past each other and the solid becomes liquid. Heat a liquid and the particles can break away completely and fly around at high speed, just as Bernoulli had envisaged (see page 23).

In 1859, Maxwell realized that with so many gas particles colliding randomly, there would be a range of speeds among them. A few would be barely moving, most would have speeds close to the average, and some would be traveling much faster. He used the mathematics of probability to work out the statistical distribution of the particles' speeds. Austrian physicist Ludwig Boltzmann

DISTRIBUTION OF PARTICLE SPEEDS AT DIFFERENT TEMPERATURES

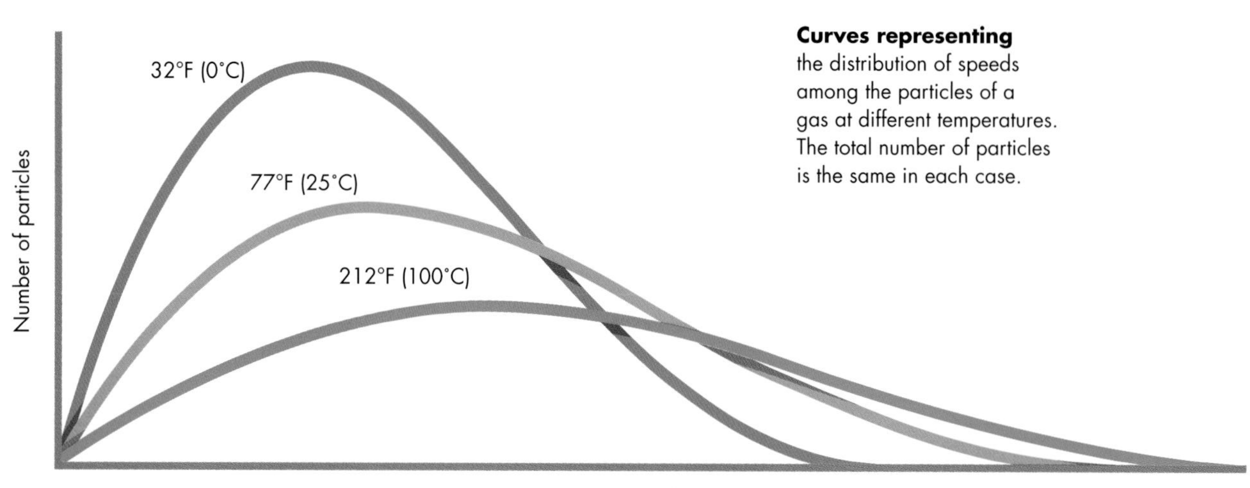

Curves representing the distribution of speeds among the particles of a gas at different temperatures. The total number of particles is the same in each case.

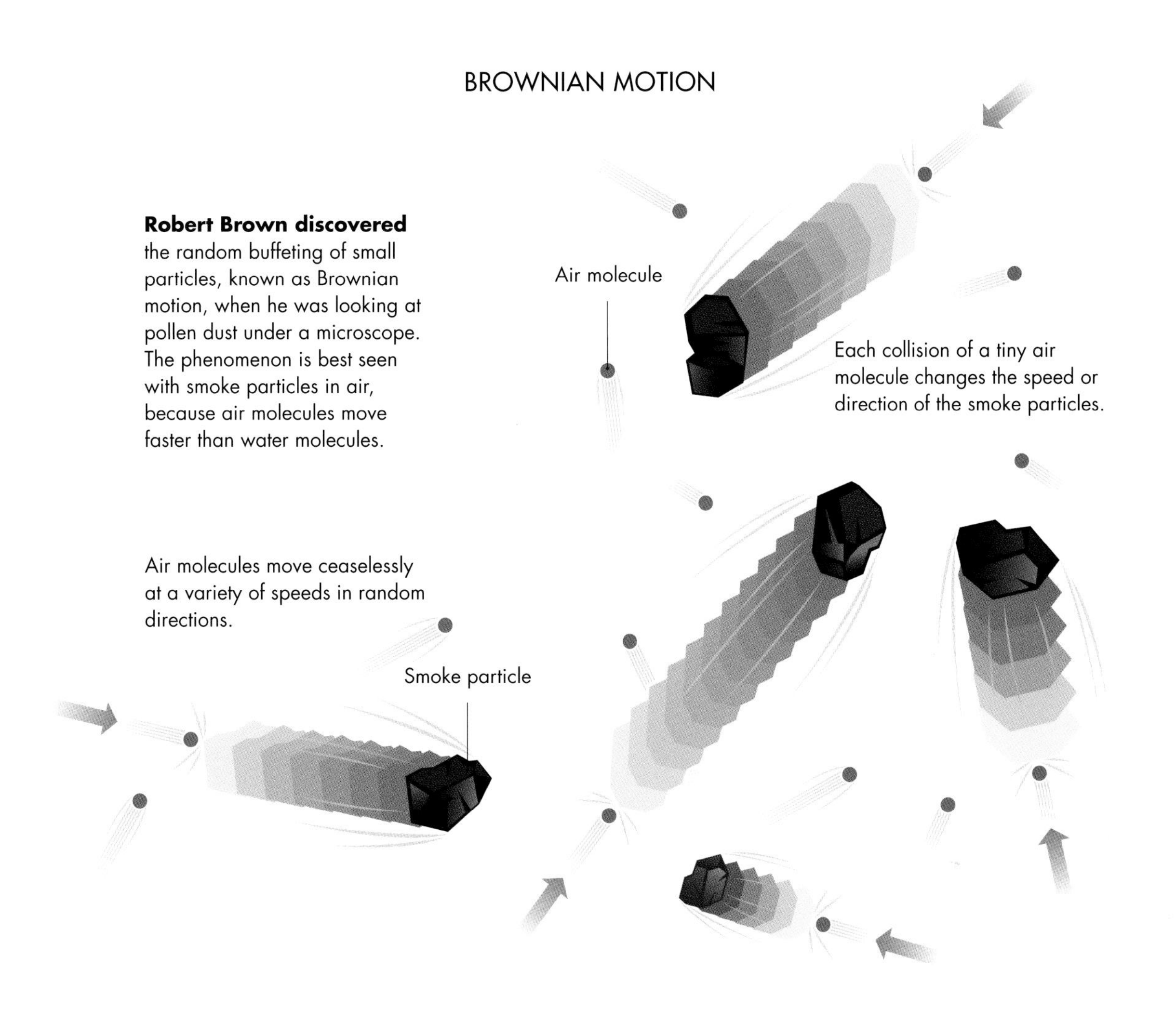

(1844–1906) improved on Maxwell's work in the 1870s. The "Maxwell-Boltzmann distribution" helped scientists to make predictions and gain new insight about gases. More importantly, Boltzmann founded "statistical mechanics"—a new branch of science that explains how the properties of countless tiny atoms can determine everyday properties of matter.

CONVINCING THE SKEPTICS

Despite the explanatory power of the Maxwell-Boltzmann distribution and the good sense that atomism made in chemistry, significant numbers of physicists and chemists still rejected atomism, thinking it was just an interesting hypothesis. The last doubters were eventually convinced in 1905, when German physicist Albert Einstein (1879–1955) derived a mathematical explanation for a phenomenon called Brownian motion. In 1827, British botanist Robert Brown (1773–1858) observed through his microscope tiny dust particles moving around in a jittery motion, as if they were being pushed around by even smaller particles that were too small to see. Einstein's mathematical analysis of Brownian motion proved that the motion of the dust particles was due to the random bombardment of water molecules. Atoms and molecules were real—and matter really was made of tiny particles. However, it was not quite that simple; by now, physicists knew that atoms were not indivisible after all.

INSIDE THE ATOM

With the existence of atoms finally confirmed, the first half of the twentieth century saw an incredibly rapid accumulation of knowledge about the atom's inner workings. Using a host of new tools and guided by innovative theories, physicists uncovered a bizarre world in which particles are waves and waves are particles—where the laws of nature challenge intuition.

ATOMIC STRUCTURE

In 1897, British physicist Joseph John (J. J.) Thomson (1856–1940) discovered the electron, a particle with negative electric charge. It was much smaller than an atom, and Thomson realized that electrons must in fact be present in every atom: "As corpuscles of this kind can be obtained from all substances, we infer that they form a constituent of the atoms of all bodies," he wrote.

Physicists knew that each atom must have equal amounts of positive and negative charges, so that it would have no charge overall. In 1904, Thomson proposed that the "negatively charged corpuscles," as he called electrons, might be arranged in rings inside a fuzzy sphere of positive charge. This idea came to be called the "plum pudding model" of the atom; the electrons were the plums and the positive charge the pudding.

One year before Thomson discovered the electron, French physicist Henri Becquerel (1852–1908) was experimenting with compounds of uranium. Becquerel discovered what Polish-born physicist Marie Curie (1867–1934) would two years later call "radioactivity." With her husband Pierre (1859–1906), Curie discovered that radiation was coming from the uranium atoms themselves and was not the result of a chemical reaction.

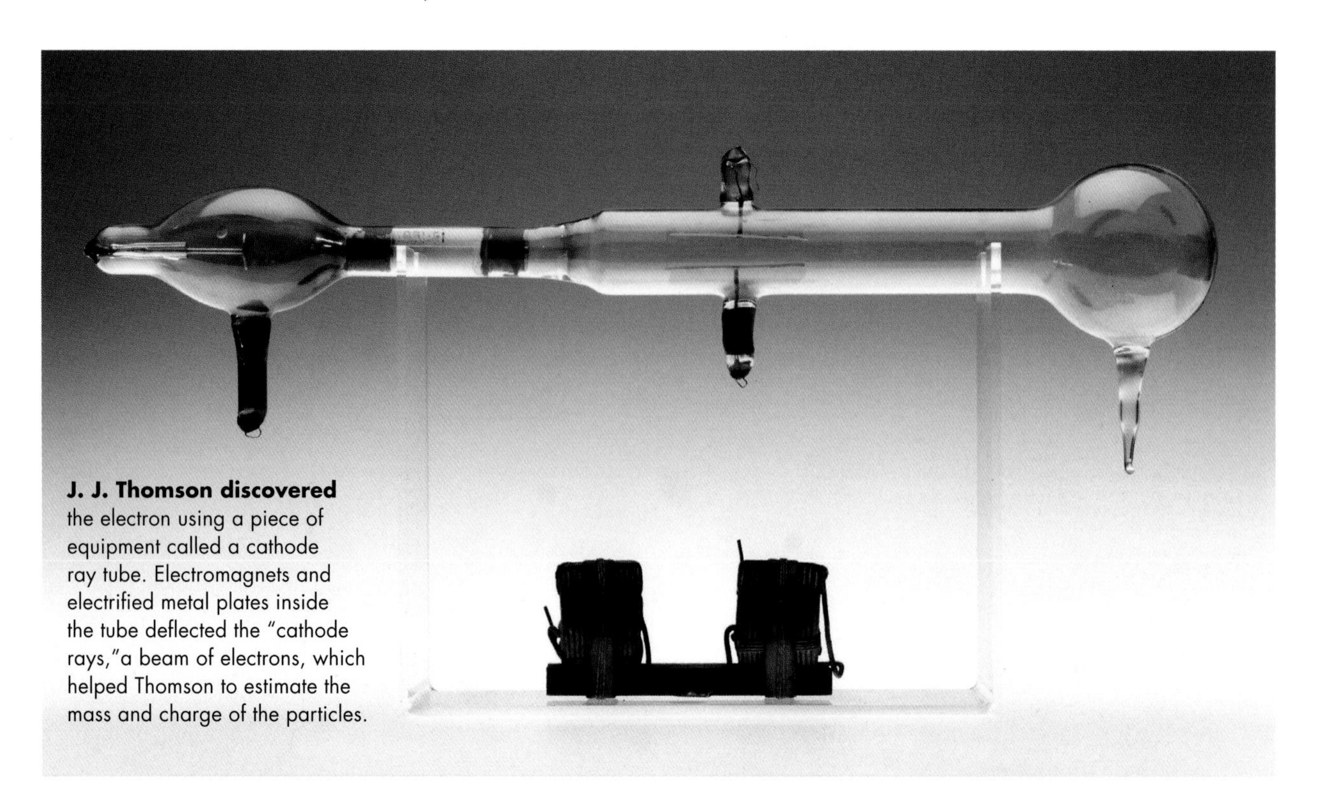

J. J. Thomson discovered the electron using a piece of equipment called a cathode ray tube. Electromagnets and electrified metal plates inside the tube deflected the "cathode rays,"a beam of electrons, which helped Thomson to estimate the mass and charge of the particles.

DISCOVERING THE NUCLEUS

To test Thomson's plum pudding model, Rutherford proposed bombarding a very thin piece of gold foil with alpha particles from a radioactive source. Measuring the deflection of the particles as they passed through the foil would enable him to work out the distribution of positive and negative charges in the atom. Rutherford's colleagues, German physicist Hans Geiger (1882–1945) and British-New Zealand physicist Ernest Marsden (1889–1970), carried out the experiment in 1911. The results were surprising. While most alpha particles passed through the foil with little or no deflection, some bounced off at more extreme angles, and about one in eight thousand bounced right back. These results were clear. The positive charge is not distributed evenly throughout the volume of the atom. Instead, it is concentrated in an extremely tiny, dense object at the center, which Rutherford named the nucleus. Rutherford later commented that it was "almost as incredible as if you fired a 15-inch shell at a piece of tissue paper and it came back and hit you."

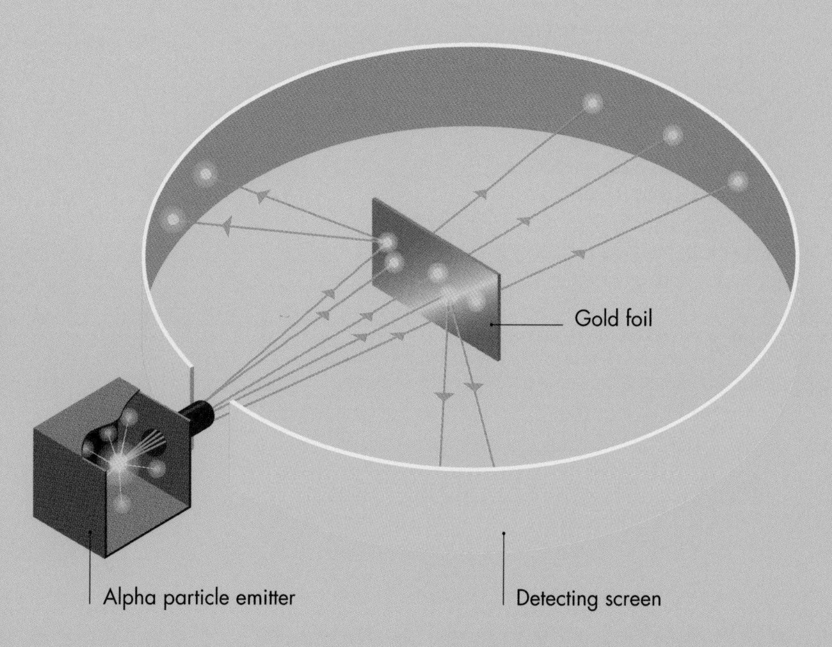

Alpha particles would pass through Thomson's atoms, with the positive charge distributed evenly. Only Rutherford's model could explain the experimental results.

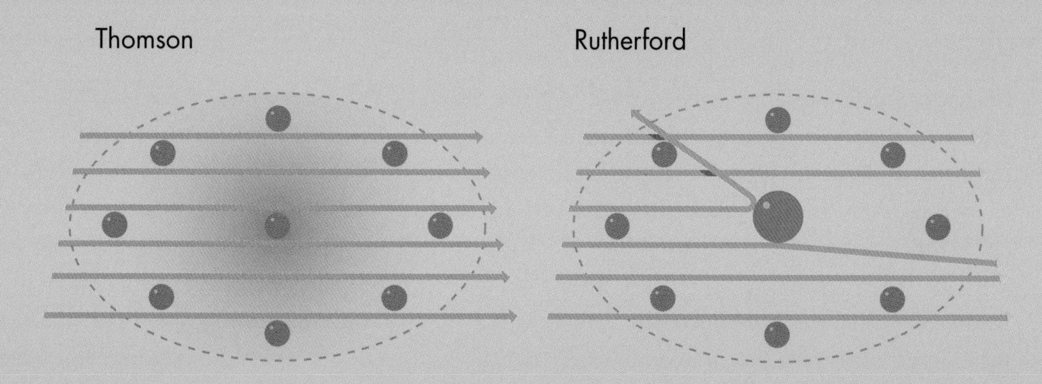

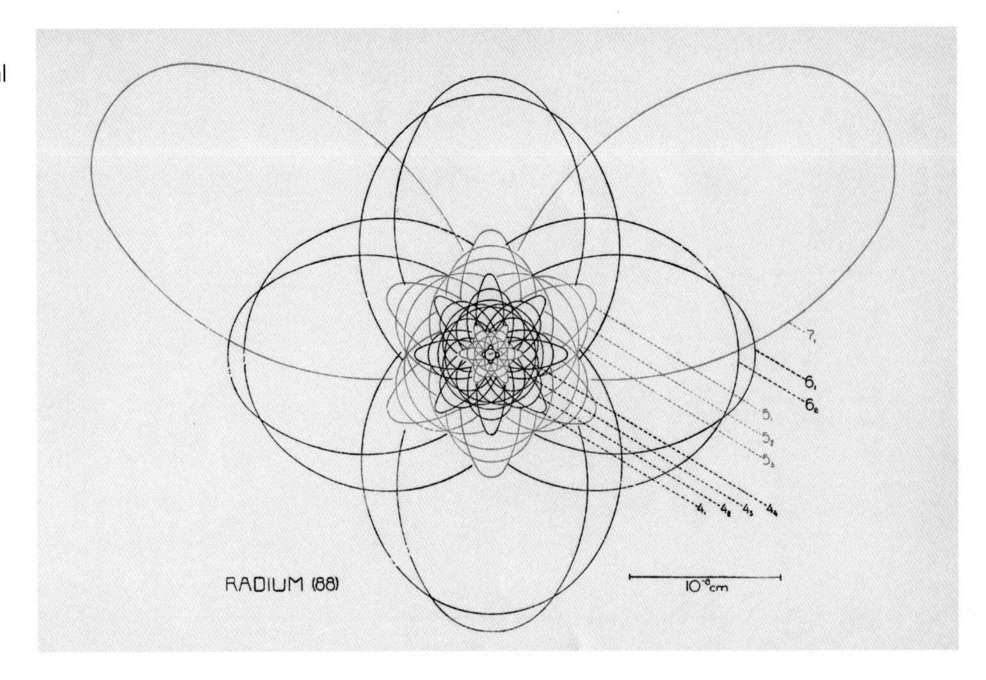

Diagram showing the atomic structure of the chemical element radium (discovered in 1898), based on the model of the atom proposed in 1913 by Danish physicist Niels Bohr and refined by German physicist Arnold Sommerfeld. This diagram was published in *The Atom and the Bohr Theory of Its Structure* (1926) by Hendrik Anthony Kramers and Helge Holst, originally published in Danish in 1922.

In 1899, New Zealand-born British physicist Ernest Rutherford (1871–1937) found that radioactive substances emit two distinct types of radiation, which he called alpha (α)and beta (β) rays. In 1900, French chemist Paul Villard (1860–1934) discovered a third type, termed gamma rays (γ). Alpha rays are streams of positively-charged particles, and they were to become especially useful tools for probing the atom. In 1911, for example, Rutherford used them to explore the distribution of electric charge in the atom (see box, page 31). The results ruled out Thomson's plum pudding model. Instead, they showed that the atom's positive charge is concentrated in an extremely tiny object at the center, which Rutherford called the nucleus.

Rutherford proposed a model of the atom in which the electrons orbit the tiny, dense nucleus, held in their orbit by the attraction between their negative charge and the positive charge of the nucleus. This is similar to how the planets of the solar system are held in their orbits by gravitational attraction to the Sun. However, there was a major problem with this model. Any orbiting object constantly changes direction, and, because changing direction constitutes an acceleration, an electron constantly radiates electromagnetic radiation (see page 28). As it does so, it loses energy, so it orbits lower and lower, spiraling inexorably down toward the nucleus.

WAVES AND PARTICLES

For a solution to Rutherford's problem, Danish physicist Niels Bohr (1885–1962) looked to a new branch of physics, quantum theory, whose effects are noticeable only at the atomic scale. According to quantum theory, only certain levels of energy are "allowed" in any given system. Because electrons' energy levels are determined by the distance at which they orbit the nucleus (the farther away, they are the more energy they have), allowed energy levels means "allowed orbits." In Bohr's (1913) model of the atom,

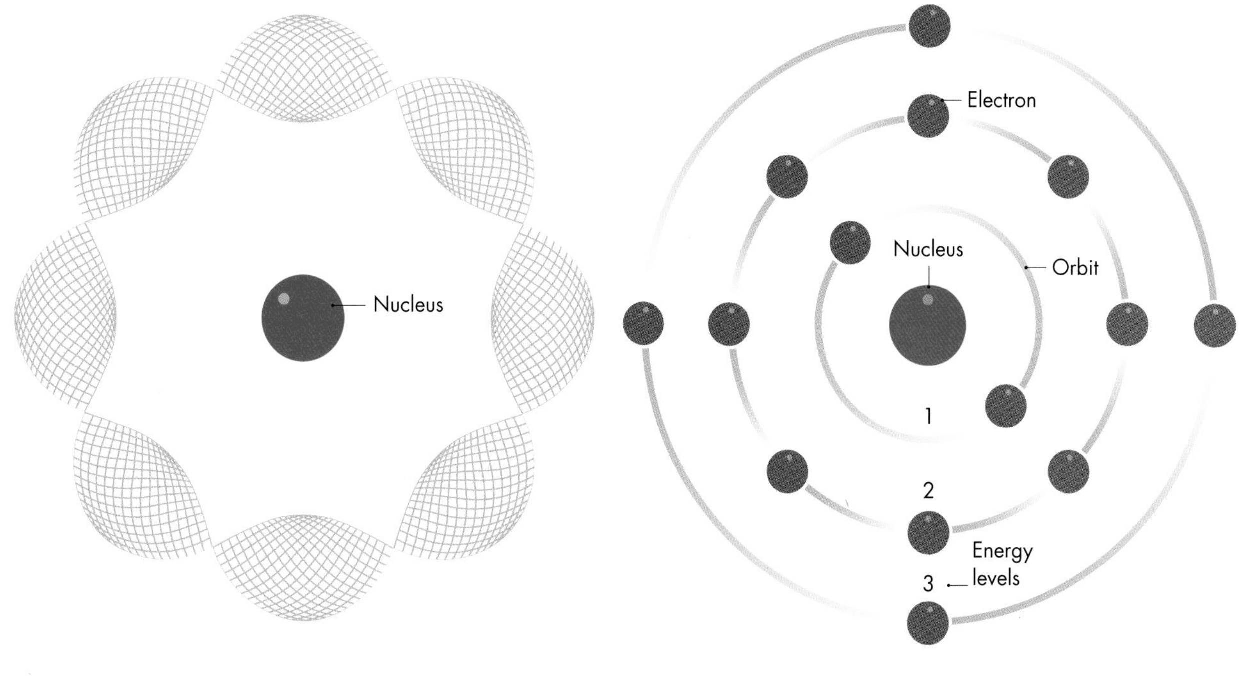

Louis de Broglie suggested that electrons are wavelike, with a wavelength corresponding to their energies. The wavelengths fitted perfectly into Bohr's allowed orbits.

Niels Bohr proposed that an atom's electrons orbit the nucleus only in specific orbits, and that electrons emit light of specific wavelengths as they "fall" from higher to lower orbits.

electrons are not able to spiral into the nucleus or continuously to emit radiation. But they are able to jump between orbits, either when they absorb energy from incoming electromagnetic radiation (to jump up a level) or by emitting electromagnetic radiation (to jump down). The frequency of the radiation depends upon the difference in energy between the two levels.

When Bohr put numbers into his model that correspond to real values, such as the electric charge on an electron, he calculated exactly the frequencies of the red, blue, and ultraviolet light that real hydrogen atoms emit. Bohr's hypothetical orbits seemed to match what spectroscopists had found (see page 68).

A few years before Bohr devised his atomic model, Einstein discovered that light and other electromagnetic radiation are formed of a stream of particles, which he named "photons," as much as the waves described by James Clerk Maxwell (see page 28). In 1924, French physicist Louis-Victor de Broglie (1892–1987) suggested that this "wave-particle duality" might also apply to entities that had always been considered as particles—in particular, to electrons. De Broglie reenvisaged Bohr's orbits, fitting "electrons as waves" around the orbits instead of a solid moving particle. Again, the numbers worked out perfectly; the distances at which the electron waves fitted exactly around the nucleus corresponded perfectly with Bohr's orbits.

SCHRÖDINGER WAVE FUNCTION

The wave function of a hydrogen atom as a graph of the probability of an electron being at a particular location, plotted against distance from the nucleus—from zero at the nucleus to a maximum, then tailing off to zero again. In three dimensions, the graph describes a hollow "cloud" of varying probability density.

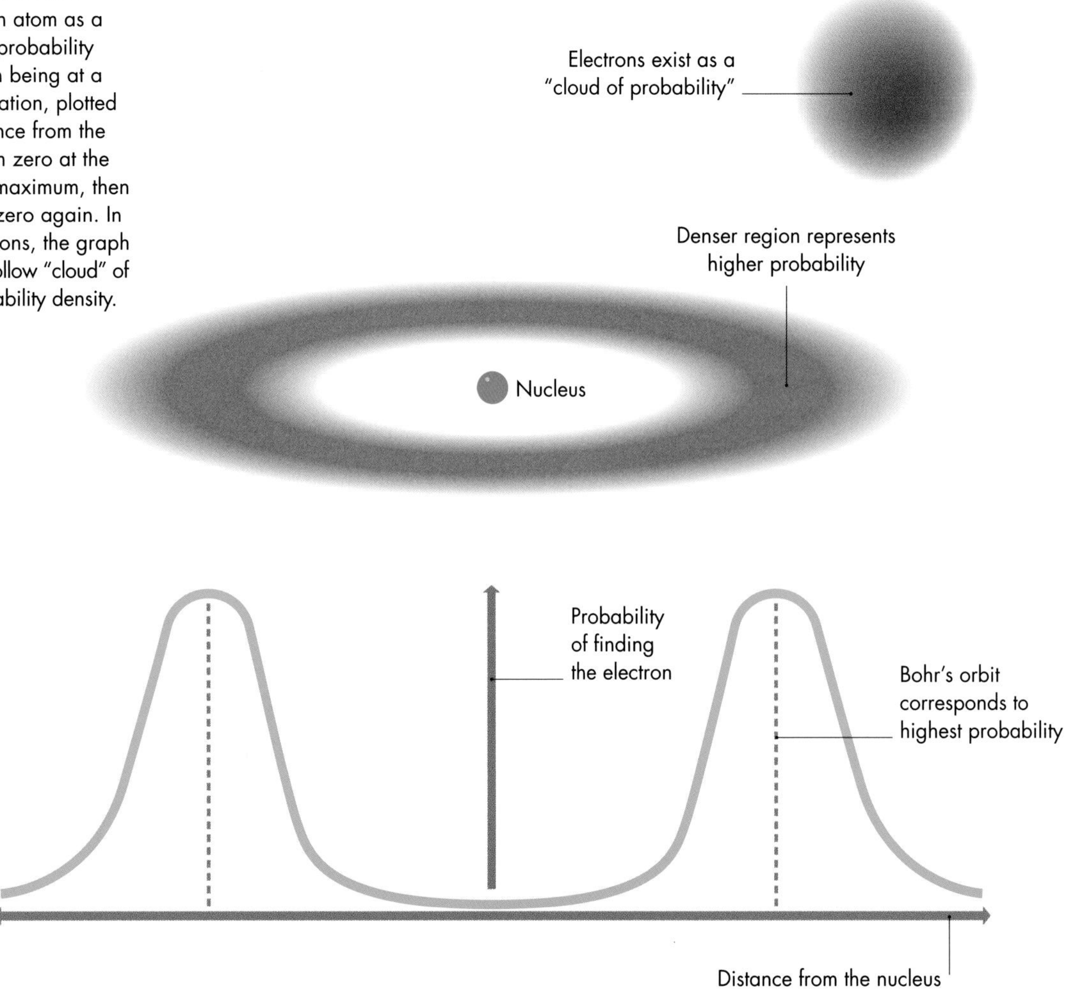

The following year, Austrian physicist Erwin Schrödinger (1887–1961) formulated an equation that allowed physicists to calculate and predict the behavior of wavelike particles (or particle-like waves) in atoms. The Schrödinger equation predicts the behavior of electrons in atoms, and in fact any quantum system, as a "wave function": a mathematical construct that determines parameters such as the position and speed of a particle. Bizarrely, perhaps, the wave function can actually give only the probability of those parameters having certain values at certain times and places. So not only does an electron behave as both a wave and particle, it exists throughout a particular region of space, as a "cloud of varying probability." We will explore the strange rules of the quantum world, including some of the later developments in understanding electrons, later in the book.

INTO THE NUCLEUS

In 1911, Dutch physicist Antonius van den Broek (1870–1926) proposed that each element has a unique "atomic number," which would be equal to the number of positive charges in the nucleus. A theory from a hundred years earlier had suggested something similar. English chemist William Prout (1785–1850) had noticed that atomic weights seemed to be exact multiples of the atomic weight of hydrogen, and suggested that the atoms of other elements might be built up as aggregations of hydrogen atoms. Prout called his hypothetical prototype atom a "protyle." In 1917, Ernest Rutherford proved that the nucleus of a hydrogen atom, with a charge of +1, is indeed found in other nuclei. Rutherford gave it the name "proton."

It was clear, however, that the nucleus must contain another type of particle, because doubling the atomic number (the amount of positive charge in the nucleus, equal to the number of protons) would more than double the atomic weight. For example, while hydrogen has atomic number 1 and atomic weight 1, helium, with atomic number 2, has a weight of 4. In 1920, Rutherford proposed that another particle must inhabit the nucleus, and suggested it is a combination of a proton plus an electron. The combination of proton (+) plus electron (-) would give this particle a neutral charge overall, so Rutherford named it "neutron." British physicist James Chadwick (1891–1974) discovered the neutron in 1932, but it was not a combination of a proton and an electron: it was a particle in its own right.

One of the big mysteries of the 1930s was just how the nucleus could hold together. After all, protons are all positively charged, and objects carrying electric charge of the same type repel. Packed so closely together in the nucleus, the forces pushing the protons apart must be very strong. Several scientists suggested that there must be an attractive force that holds together the protons and neutrons inside the nucleus. Because neutrons contribute to the attractive force but not the repulsive one (because they have no electric charge), they act as a kind of glue to hold the nucleus together.

In 1935, Japanese theoretical physicist Hideki Yukawa (1907–81) suggested a mechanism by which this "strong nuclear force" might work. He proposed that protons and neutrons constantly pass between them mediating particles called "mesons"—and that the exchange of these particles is the source of the force. It was a bold theory but, nevertheless, Yukawa's mesons were discovered in 1947. We will explore the nucleus, including some of the later experimental and theoretical developments, in later chapters.

A DEEPER ATOMISM

Yukawa's meson was a member of a burgeoning family of tiny particles, all much smaller than atoms. These particles were being discovered in cosmic rays (streams of particles coming from space) and in particle accelerators. In 1964, American physicist Murray Gell-Mann (born 1929) suggested that some subatomic particles are "composite"—made up of yet smaller particles, called quarks. According to Gell-Mann's theory, protons and neutrons are composed of three quarks, while other particles, such as mesons, are composed of two. Other particles, such as the electron, are truly fundamental. The current best theory for understanding the plethora of fundamental and not-so-fundamental particles—the bastion of modern atomism—is the Standard Model, which we explore in chapter seven.

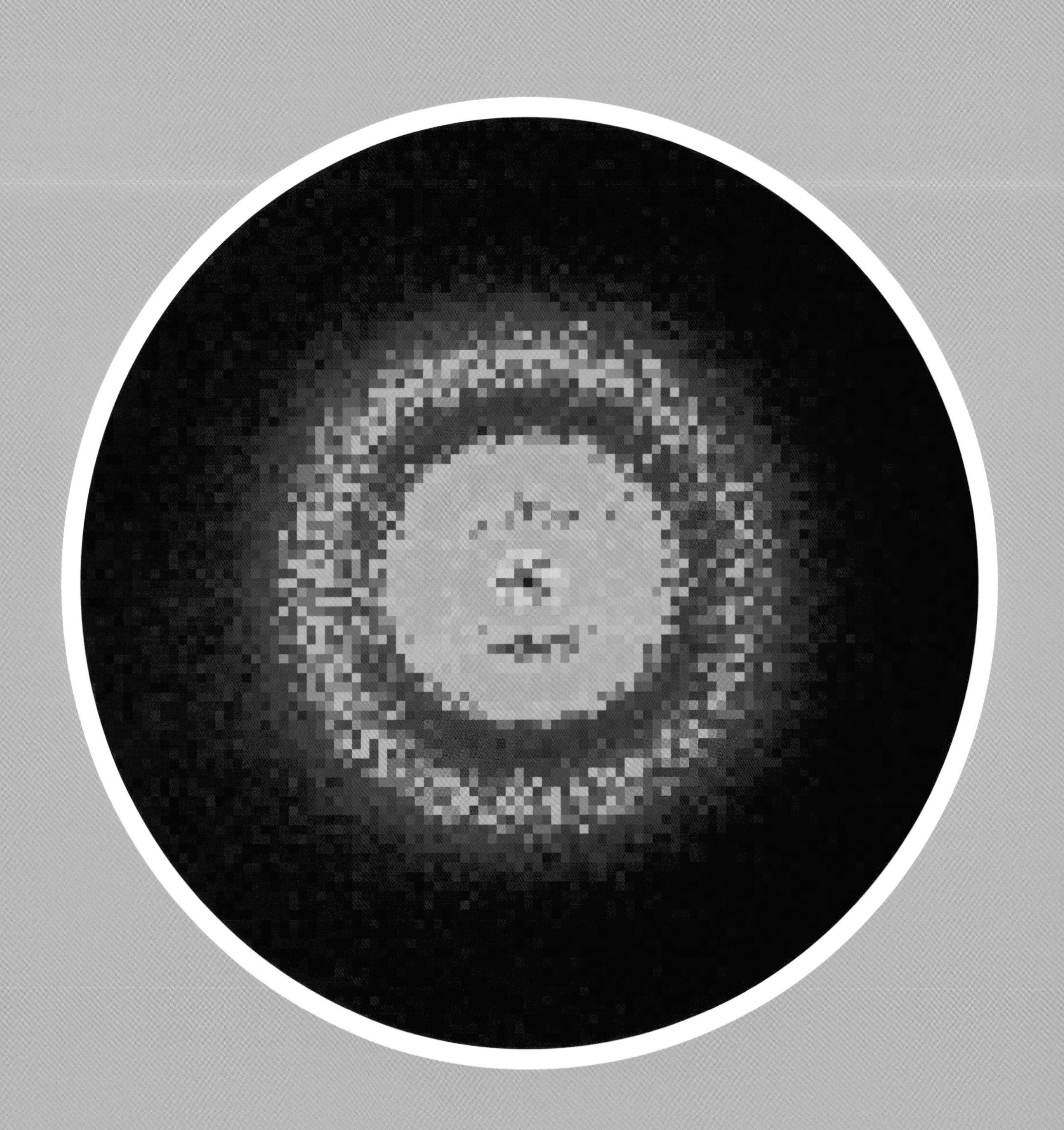

CHAPTER 2
STRUCTURE OF THE ATOM

An atom is made of just three types of particle: protons, neutrons, and electrons. The protons and neutrons cling tightly together to form the incredibly dense nucleus at the atom's center. Surrounding the nucleus, held in place by electrical attraction, are the electrons. These three types of particle behave according to the strange laws that apply at the atomic scale—laws that are codified within a branch of science known as quantum mechanics.

Although electrons are particles, they also have wavelike properties. Most of the time they exist in many places at once, as a "cloud of probability" called a wave function. This image, captured with a "quantum microscope," shows the electron wave function of a single hydrogen atom.

A SENSE OF SCALE

Atoms are unimaginably small. You would need a line of about 250 million of them to reach one inch, or ten million of them to reach one millimeter. They are so small and weigh so little that even tiny objects are made of countless trillions of them.

GETTING THE MEASURE OF ATOMS

With diameters of around four billionths of an inch, or one ten-millionth of a millimeter (see box), atoms are extremely small. We can, however, gain an interesting and useful perspective on the size of an atom if we scale it up to everyday dimensions. Scale up by a factor of ten million and the atom is now the size of a typical grain of fine sand (0.04 inch, 1 millimeter). If we scale up everyday objects by the same factor (as if they are made of atoms the size of small sand grains) then a soccer ball is now two-thirds the size of the Moon; a common housefly is 44 miles (70 kilometers) long; and the period at the end of this sentence is 3 miles (5 kilometers) in diameter.

The atomic nucleus—the dense central part of an atom—constitutes more than 99.9 percent of the atom's mass, but it takes up only a tiny proportion of the space. Its diameter is about one hundred-thousandth (10^{-5}) the diameter of the whole atom, so it takes up only one thousand-trillionth (10^{-15}) of the atom's volume. We need to scale up many more times to make the nucleus big enough to be visible. Enlarge the nucleus so that now it is about the size of a grain of sand, and the now-even-more-enlarged atom is about 110 yards (100 meters) in diameter—the length of a football field. At that scale, the period at the end of this sentence would now cover most of the Moon's orbit around Earth.

ATOMIC SCALE UNITS

Scientists use the International System of Units (SI). Length measurements are based on the metric system, so they use meters and centimeters, instead of feet and inches. In the metric system, small everyday objects are typically measured in centimeters (cm, hundredths of a meter, 10^{-2}m) or millimeters (mm, thousandths of a meter, 10^{-3}m). One centimeter is about 0.4 inch; one millimeter is about 0.04 inch. At the atomic scale, the most commonly used units are nanometers (nm, billionths of a meter, 10^{-9}m) and picometers (pm, trillionths of a meter, 10^{-12}m). A typical atom has a diameter of a few ten-billionths of a meter (10^{-10}m). This, annoyingly, is "a few tenths of a nanometer" or "several hundred picometers." However, there is one unit that is perfectly suited to the size of atoms and has a special place in the hearts of many atomic physicists: the ångström. One ångström is one ten-billionth of a meter (1Å = 10^{-10}m), so atoms are a few ångströms in diameter. In fact, the diameter of the smallest atom, an atom of hydrogen, is almost exactly 1Å, while the biggest atoms, atoms of the element cesium, are about 6Å in diameter. It is worth noting that there are several ways to measure or calculate the diameter of an atom—and, in fact, atoms do not really have a definite diameter, as we will see.

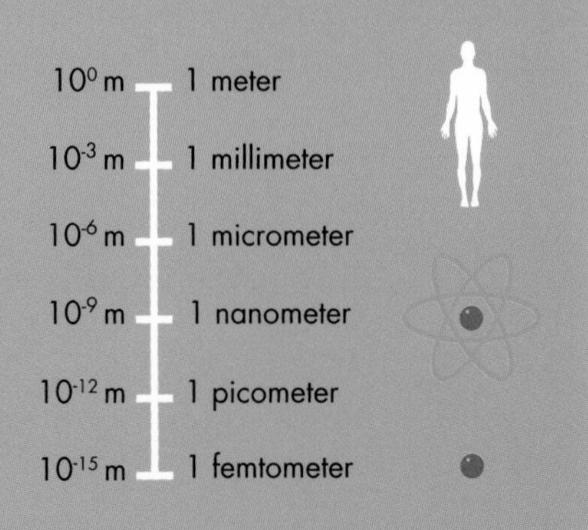

A housefly is made of about a hundred billion billion atoms. If each of those atoms was the size of a grain of sand, the fly would be so huge it would cast a shadow over most of Long Island, New York. Using the same analogy, with atoms the size of a grain of sand, a soccer ball would have a diameter of nearly 1,400 miles (2,250 kilometers), approaching two-thirds the Moon's diameter of 2,160 miles (3,476 kilometers).

WEIGHING IT UP

Another way to appreciate how tiny an atom is, involves considering its mass. The mass of the electron is less than one eighteen-hundredth the mass of a proton or neutron, and therefore negligible—so the mass of an atom is determined by its total number of protons and neutrons, collectively known as "nucleons." Protons and neutrons have almost exactly the same mass. At the atomic scale, mass is usually measured in "unified atomic mass units" (u), also known as "daltons" (Da). One dalton is approximately the mass of one nucleon, though the precise definition is that 1 Da is equal to one-twelfth of the mass of a carbon-12 atom (an atom that contains twelve nucleons: six protons and six neutrons). The mass of an isolated proton or neutron is very slightly more than 1 Da—see facing page.

Close to 99 percent of carbon atoms are carbon-12. Nearly all the others are carbon-13 atoms, which have six protons and seven neutrons (a total of 13 nucleons, so the mass of a carbon-13 atom is about 13 Da). Because carbon-12 is by far the most common, the average mass per carbon atom is very close to 12 Da: in fact, it is 12.01 Da.

To appreciate how tiny these amounts of mass are, consider the difference between daltons and ounces. To convert daltons into ounces, you need to multiply by 17 trillion trillon (17,000,000,000,000,000,000,000,000. So, if you have that many nucleons, they would have a total mass of 1 ounce (28 grams), while that many carbon atoms will have a mass of 12.01 ounces (340 grams).

Scientists work in grams and kilograms, instead of ounces. The number you need to convert daltons into grams is six hundred thousand billion billion (600,000,000,000,000,000,000,000). That number of atoms of any element will have a mass equal to the atomic mass in grams. So six hundred thousand billion billion helium-4 atoms , with four nucleons, will weigh 4 grams (about ⅛ ounce), while the same number of uranium-238 atoms, with 238 nucleons, will weigh 238 grams (about 8⅜ ounces). This useful number is called Avogadro's constant, in honor of Italian scientist Amadeo Avogadro (1776–1856). In the 1810s, Avogadro developed the notion that identical volumes of any two gases at the same temperature and pressure would contain the same number of particles (atoms or molecules), although he had no way of knowing how many there might be.

Accurate calculations of Avogadro's constant were first made early in the twentieth century. The number defines what chemists call a "mole" of a substance, which is useful in tracking chemical reactions. One mole of a substance always contains a number of particles equal to the Avogadro constant. So, for example, to make one mole of water (18 grams or ⅝ ounce), an amount that

In 12 grams (⅜ ounce) of pure carbon there are about six hundred thousand billion billion carbon atoms.

ATOMIC MASS UNITS

The protons and neutrons that make up an atomic nucleus are bound together by strong forces, so it would take energy to break them apart. As a result, a nucleus has less energy than the energy those protons and neutrons would have if they were separate from each other. The difference is called binding energy. Energy has mass, as attested by Albert Einstein's most famous equation, $E = mc^2$, so, a nucleus also has less mass than the protons and neutrons of which it is made. The difference is called the mass defect, and it is different for different nuclei. In order to have a standard way of expressing the mass of a nucleus, then, scientists use a unit called the unified atomic mass unit, or dalton, which is defined as one-twelfth the mass of the carbon-12 nucleus.

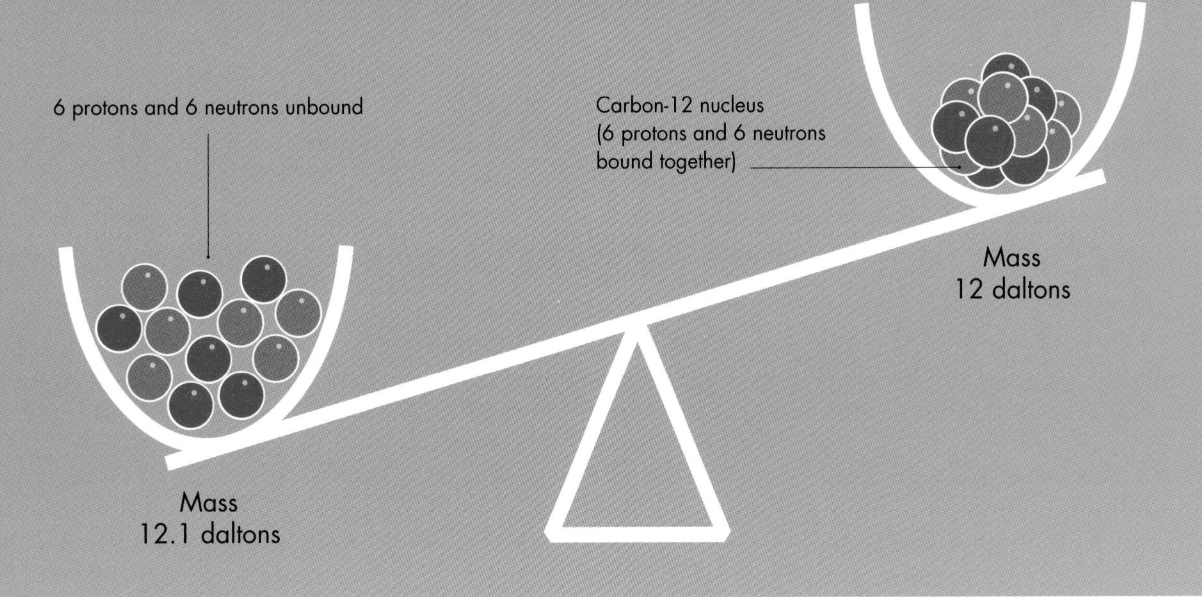

would fill just one tablespoon, you would need two moles of hydrogen (2 grams or 7⁄100 ounce) and one mole of oxygen (16 grams or about ½ ounce). The number of water molecules is equal to Avogadro's constant, because there is one mole of it, but the ingredient atoms number three times as many altogether.

Incidentally, Avogadro was right about gases; identical volumes of any two gases really do contain the same number of particles, as long as they are at the same temperature and pressure. One mole of any gas at normal atmospheric pressure and a temperature of 32° F (0° C) will take up 4.9 gallons (22.4 liters). This is about the space inside a typical domestic microwave oven. Next time you look in through a microwave door, imagine the air inside being composed of hundreds of thousands of billions of billions of incredibly tiny atoms and molecules, dashing around at high speed.

WHAT ATOMS ARE LIKE

Considering the inconceivably small size of atoms, it is remarkable that we know about their existence at all, let alone anything about their internal structure. And yet, we know a great deal. The traditional picture of an atom, with its electrons following well-defined paths around the nucleus, is in fact far from correct.

ELECTRON ATMOSPHERE

As soon as scientists realized that atoms are not solid, impenetrable balls—that they must possess inner structure—they began to hypothesize about how the various parts might be organized (see page 30).

If you could actually "see" an isolated atom, it would look like a fuzzy ball. The electrons form something like a spherical cloud surrounding and obscuring the nucleus, which would in fact be far too small to see. Just like the atmosphere around a planet, the electron cloud does not have a hard edge or a definite boundary; instead, it peters out gradually. In all except hydrogen and helium atoms, there are two or more fuzzy spheres, arranged concentrically. Electrons in the outermost spheres have the most energy. When atoms are not isolated but are instead joined together, their electron clouds can have different shapes, and they can overlap. This is how atoms bond to make molecules, as we will see in chapter four.

The fuzziness of the spherical electron cloud is not the result of the motion blur of fast-moving electrons. Instead, it is because of something uncanny in the nature of everything at the atomic and subatomic scale—the fact that, at that scale, particles have wavelike qualities. This strange "wave-particle duality" dominates the behavior of atoms and their component parts. It is one of the most important facets of a branch of physics known as quantum mechanics.

CLASSIC MODEL OF THE ATOM

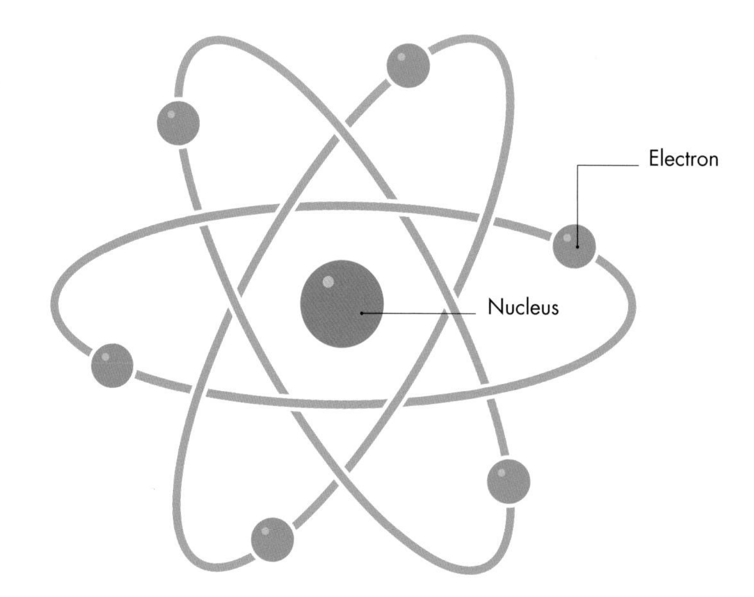

The most enduring image of the atom is one in which the electrons orbit the nucleus, much as planets orbit the Sun, as if each atom is a tiny solar system. So appealing is this model of the atom that it remains iconic to this day. However, it is a far-from accurate picture of what an atom is like.

MODERN PICTURE OF AN ATOM

Illustration of a carbon-12 atom, showing the fuzzy electron clouds. In the nucleus are six protons and six neutrons—a total of twelve nucleons. The nucleons are composite particles, made of different combinations of even smaller particles called "quarks." There are several different types of quark. Protons and neutrons are made of two types, which physicists have named "up" and "down." The up quark has an electric charge of +2/3, while the down quark has a charge of –1/3. With two ups and a down, the overall charge on a proton is +1; with two downs and an up, the neutron has no overall charge.

Each of the six electrons carries a charge of –1, so they are held in place around the nucleus, by electrostatic attraction between them and the protons. The electrons do not follow strict paths around the nucleus—instead, they have a probability of being in many places at the same time. So the best way to think of the electrons around an atom is as "probability clouds." The electrons of an isolated carbon atom are in two concentric clouds, both spherically symmetrical. The electrons in the outermost cloud have more energy than those in the innermost cloud.

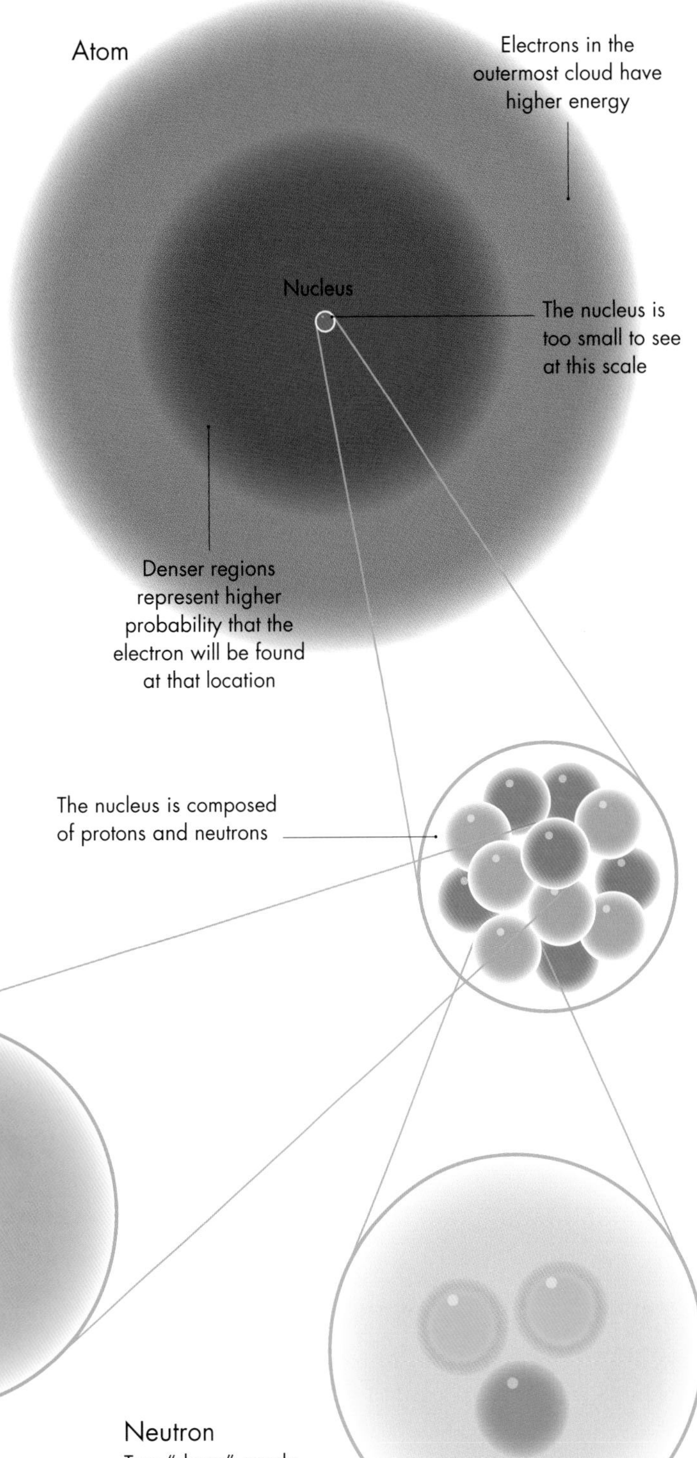

QUANTUM MECHANICS

NEWTON'S LAWS

Newton set out his three laws of motion in his groundbreaking book *Philosophiæ Naturalis Principia Mathematica* (*Mathematical Principles of Natural Philosophy*, 1687). Simply put, and paraphrased unmathematically, the first Law states that an object's motion will not change if no force acts on it. The second law states that an object will accelerate if a force does act, changing its speed and/or direction of motion by an amount that depends upon the strength of the force and the mass of the object. The third law states that whenever one object exerts a force on another, the second object exerts an equal force, in the opposite direction, on the first object. In the same book, Newton used his laws of motion and his understanding of gravity to construct a thought experiment involving a cannon fired horizontally from the top of a mountain (see below). The cannonball will fall to the ground—the faster it shoots out of the cannon, the farther it travels before hitting the ground. If it is fired fast enough, its fall will match the curvature of Earth, and it will go into orbit.

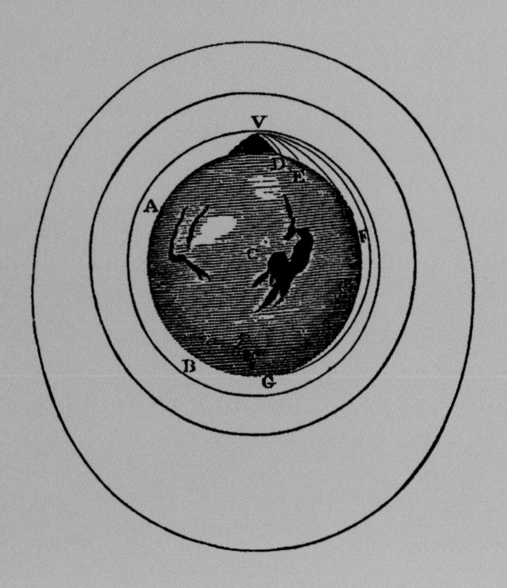

A reality in which electrons can exist in many places at once, in which they can have particle-like and wavelike properties, and in which their behavior is guided by probability, is unfamiliar. These strange goings-on at the atomic scale can be predicted by the most robust and well-tested area of modern science: quantum mechanics.

WHAT IS MECHANICS?

Mechanics is a branch of physics with which one can predict the behavior of objects subjected to forces. It is possible, for example, to determine how long it will take for a ball thrown in the air to come back down and where it will land—as long as you know its mass, how fast and in what direction it was thrown, and any forces acting on it, such as gravity and air resistance. The pioneer of mechanics was Isaac Newton, who set out his famous three laws of motion in 1687 (see the box). Expressed mathematically, these laws were all that was needed to land people on the Moon. Architects and engineers also use them routinely, for example, when designing bridges or engines.

Despite the success and widespread applicability of Newton's laws, physicists in the second half of the nineteenth century found them to be a flawed description of reality. Challenges to Newton's cozy "mechanistic" vision of the universe led to the founding of modern physics in the early twentieth century. The first challenge came from James Clerk Maxwell's elegant theory of electromagnetism (see page 28) —the theory that led to the realization that light is a form of electromagnetic radiation. Maxwell's theory demanded that the speed of light is "absolute"; in other words, light will always approach at the same speed, even if you are moving toward or away from its source. If two people are in motion relative to each other, a beam of light will have the same speed relative to each of them.

Newton's Laws of Motion may be flawed, but they were good enough to land people safely on the Moon.

This fact challenged the assumption of Newton (and everyone else) that time and space must be absolute—and it shook physics to its core. In the hands of Albert Einstein, it led to the inevitable conclusion that while (and because) the speed of light is absolute, distances and time intervals are "relative." People in motion relative to each other can measure the same intervals of time and space differently.

Einstein's theories of Special Relativity (1905) and General Relativity (1915) represent a reframing of mechanics to take into account the findings of Maxwell's electromagnetism. The strange effects predicted by relativity—of time running at different rates and distances being shortened or extended—are only significant at particularly high relative speeds, or in very intense gravitational fields. In these situations, physicists employ "relativistic mechanics" instead of mechanics based on Newton's laws, which are now called "classical mechanics." Incidentally, Einstein's most famous equation, $E = mc^2$, is a direct result of his theory of Special Relativity.

The mechanics that can be used to predict the behavior of objects at the atomic scale—quantum mechanics—is another nonclassical version of mechanics. It, too, came about as a result of challenges to Newtonian mechanics, and once again, Einstein played a key role in its development.

WHAT IS A QUANTUM?

The roots of quantum mechanics lie in a scientific paper published in 1900 by German physicist Max Planck (1858–1947). Planck was trying to make sense of the light (and other electromagnetic radiation) that hot objects radiate, such as the orange glow of hot coals, the white light given off by the surface of the Sun and the (invisible) infrared radiation emitted by all objects at room temperature. James Clerk Maxwell had shown that light is an electromagnetic wave, an undulation in the electromagnetic field. All electromagnetic waves are the same except for their frequency (the number of complete cycles of the undulation that are completed each second) and therefore also their wavelength (the distance between successive peaks).

Scientists also knew that the hotter an object is, the more energy it will emit as electromagnetic waves overall—and the greater will be the proportion of higher-frequency waves (waves with a shorter wavelength). So, if you heat an iron bar to a temperature of about 1,100°F (600°C), it will emit infrared radiation and some light from the lower-frequency red end of the spectrum, but virtually no light from the higher-frequency blue end. Heat it further and it will begin to produce those higher frequencies—and at a high enough temperature it will glow white-hot as its emanations cover the whole spectrum. Planck was trying to develop an equation that could predict the intensity of any frequency at any temperature.

According to Maxwell's theory, electromagnetic radiation is produced by electrically-charged objects vibrating, or oscillating. Planck did not suggest what the vibrating objects might be, but it made sense, even in terms of classical mechanics, that the energy of each of these "oscillators" is directly related to the frequency at which it is vibrating. Planck could not quite make his equation reflect reality. In what he called an "act of desperation," he proposed that an oscillator is not "allowed" to give out just any amount of energy; instead, it can only give out energy in tiny, discrete amounts that he called "energy elements."

Planck also introduced a number, which has since been called Planck's constant, h. The smallest amount of energy a particular oscillator could emit is h multiplied by the frequency of oscillation, f. In other words, $E = hf$. The oscillator can emit any number of whole energy elements, but simply is not able to emit fractions of an energy element. It's like being able to spend any number of dollars, but not cents or fractions of cents. If you are buying something that costs millions or billions of dollars, you do not notice the restriction. Similarly, Planck's energy elements disappear into insignificance at the everyday scale, because h is a very tiny number. Planck's idea represents a kind of atomism of energy.

The wavelength of violet light at one end of the visible spectrum is about half the wavelength of red light at the other end.

Light from hot objects
The range of wavelengths of light emitted by a hot horseshoe is represented on the graph as the orange curve. The curve peaks at a fairly short wavelength (red light) and has a long tail of longer wavelength infrared. Heated further, the horseshoe would produce more light overall—the green curve—and would peak in the shorter yellow wavelengths. It would glow yellow-hot. Much hotter still, and the horseshoe would produce even more radiation overall—the blue curve—and would emit short wavelength blue light, too. All the wavelengths of the visible spectrum would now be produced, and the horseshoe would be glowing white-hot.

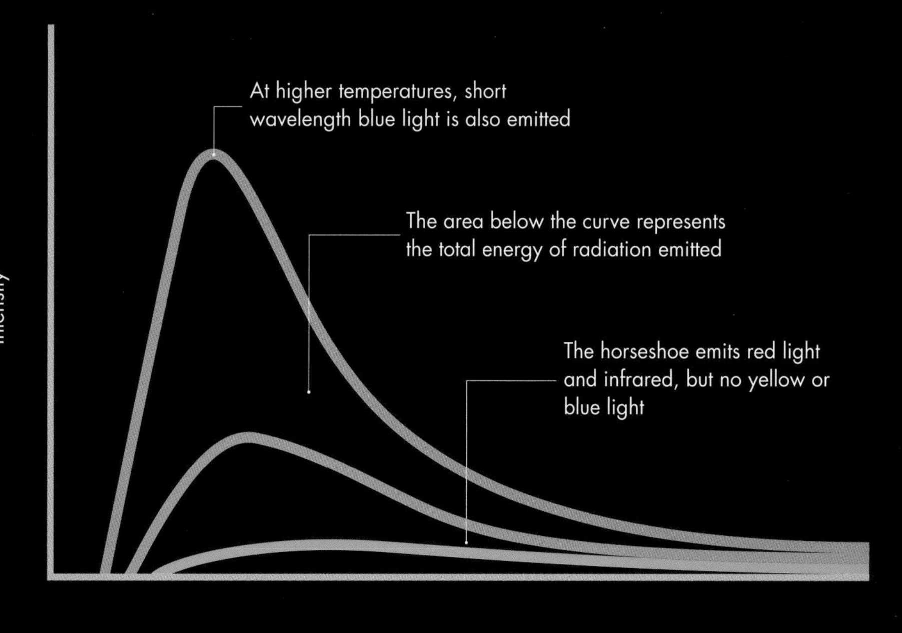

To Planck, it was just a mathematical convenience, but in the hands of Einstein, it was the beginning of a revolution in physics. In 1905, Einstein borrowed Planck's ideas to explain a phenomenon known as the photoelectric effect, in which light displaces electrons from their atoms. Einstein realized that the light's energy is delivered in discrete amounts—packets of energy, particles even, which he called "photons." Einstein used Planck's equation $E = hf$ to work out the energy of each individual photon of light. Each photon is the equivalent of one energy element, or quantum. Because violet light has about twice the frequency of red light, each photon of violet light has about twice the energy of a photon of red light. Light energy can be delivered in photons, but not in half photons. Once again, this restriction means nothing in everyday scenarios. If you shine a flashlight or light a candle, countless trillions of photons will emerge, with a range of individual energies.

The first model of the atom to take account of quantization was published in 1913 by Niels Bohr. It was long known that when you "excite" an atom, by heating it or by irradiating it with ultraviolet light, the atom will emit electromagnetic radiation with precise frequencies. This is how fluorescent dyes and paints work—and why they produce such vivid and consistent colors. Furthermore, because the precise frequencies of radiation are characteristic of the elements present, fluorescence gives chemists a way to identify elements, using spectroscopy (see page 68).

THE PHOTOELECTRIC EFFECT

The phenomenon known as the photoelectric effect involves light being shone onto a metal surface. The energy of the light displaces and frees some electrons from their atoms, and experimenters can detect the freed electrons as an electric current. The more intense the light, the more electrons—and the greater the current. Intriguingly, below a certain frequency no electrons are ejected at all, however intense the light. Einstein's (correct) explanation of this was that the energy is delivered in "chunks," particles he called photons. Each photon has a certain amount of energy that is determined by the light's frequency. The intensity of the light would then be related to the number of photons arriving at any point each second. If none of the photons has enough energy to unseat an electron—in other words, if the frequency is not high enough—then it does not matter how many photons arrive per second (that is, how intense the light is); no electrons will be ejected, and no electric current will flow. Using light with higher frequency (smaller wavelength) means each photon has more energy, and each ejected electron has more energy.

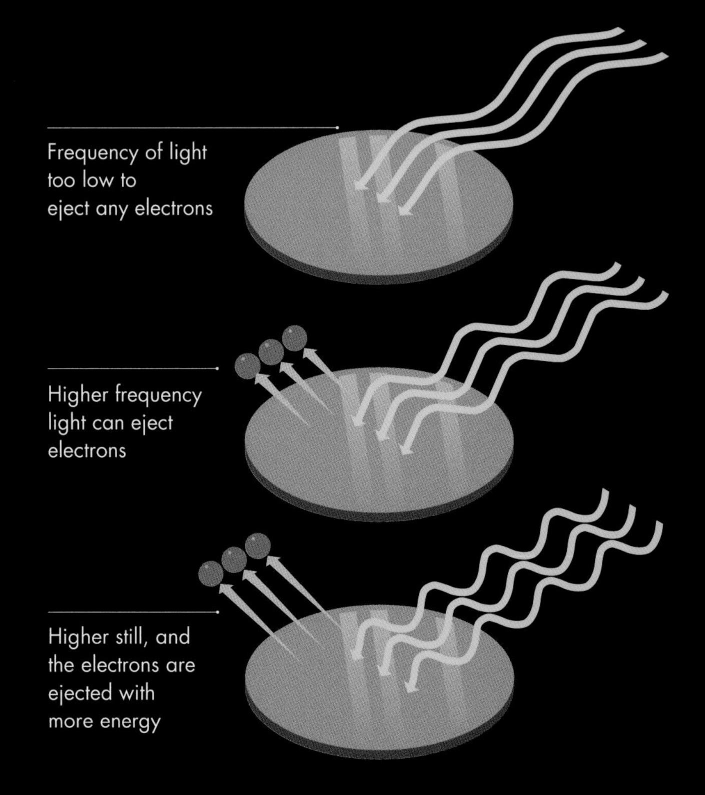

HOW ATOMS EMIT PHOTONS

In Bohr's model of the atom, electrons can only have certain "allowed" energies. The higher the energy, the farther the electron is from the nucleus. When an electron is at a higher-than-normal energy level, it can fall back down, and when it does so, it emits a photon of light. The energy lost by the electron determines the wavelength of the light.

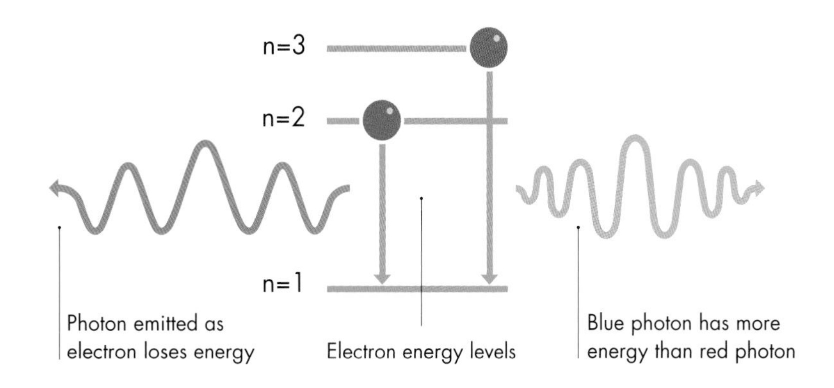

Bohr realized that because the light has precise frequencies, the photons being emitted must have definite energies. And, because it is the electrons that produce the photons, he concluded that the electrons must be jumping between definite energy levels within the atom. Otherwise, the photons would have a continuous range of energies, and the light would have a continuous range of frequencies. Bohr recognized that the whole system, including the energies of the electrons in an atom, must be quantized. He was right, and today physicists assign a number to each energy level that they term the "principal quantum number," n. The electrons with the lowest energy, $n = 1$, are said to be in the ground state.

Bohr used his theory to calculate the electron energy levels for the simplest atom, hydrogen. He then calculated the frequencies of the light that would be emitted whenever an electron jumped down from a higher-energy level to a lower one. The results perfectly matched the frequencies of the light emitted by hydrogen in the real world. It was clear that electrons in atoms really cannot have just any values, nor can they gain or lose energy continuously. Instead, they do so with sudden, tiny "quantum jumps" from one value to another. Quantization is a fact of life, albeit one that only becomes significant at particularly small scales.

Electrons in this glow-in-the-dark species of coral are being excited to higher energy levels by photons of invisible ultraviolet light, then losing energy as photons of visible light.

WAVE-PARTICLE DUALITY

The quantization of energy is one of the pillars of quantum mechanics. Another is wave-particle duality, the notion that the entities subject to the laws of quantum mechanics behave both as particles and waves. Einstein realized that light behaves as a stream of particles (photons). However, it is also clear from Maxwell's work on electromagnetism (and a host of practical experiments) that light also behaves as a wave. This duality presents a serious challenge to common sense and to everyday experience: waves are spread out and continuous, while particles are localized and discrete. To this day, no one quite understands how anything can truly be both.

To appreciate how wave-particle duality challenges our experience of the world, consider the following scenario. A light source shines onto a screen, which is covered with an array of detectors. First, consider light as a wave. It is a disturbance in the electromagnetic field, a disturbance that spreads out from the source. The farther the wave travels from the source, the more its energy is spread out, so each detector picks up a small proportion of the light's energy. All the detectors receive the same amount at the same time.

Now consider the light being delivered in discrete packets: photons. Each photon is a particle, and travels in a straight line from the source to just one of the detectors. If you send one photon at a time ("single-photon light sources" do exist), then one, and only one, of the detectors is activated. That single detector collects the whole energy of the photon; the energy is not spread out as before. Each detector has an equal chance of receiving the photon, so when large numbers of individual photons are emitted, over time, all the detectors will receive their fair share of the total energy given out by the light source. In other words, the process is determined by probability, like flipping a coin. When you flip a coin many times, each individual flipping event is independent of all the others, yet over a large number of flips, exactly half will land heads up and the other half heads down. That coins can do this is strange enough, but how individual photons can be "guided" by probability in this way remains a deep mystery.

This scenario becomes stranger still when you consider a classic quantum conundrum: the "double slit experiment." Its bizarre implications are key to quantum mechanics and therefore to understanding atoms. As its name suggests, the double slit experiment involves shining light through two narrow slits. The light falls on a screen that is covered with detectors, as before. When a light wave passes through a slit, it spreads out, as any wave does when it passes through a small opening (the result of a phenomenon called diffraction). The slit therefore behaves as a source of light whose waves radiate outward beyond it and toward the screen. With two slits, there are effectively two sources of light next to each other, each one radiating outward toward the screen. The two sources of light are identical and the waves are in step.

Where the light hits the screen, there will be places where the peaks and troughs of the light waves arrive together—where the two light waves are "in phase." At these locations, the waves reinforce each other, making a patch of light that is brighter than it would have been without the slits. Similarly, there will be places where a peak always arrives with a trough, that is, where the two waves are "out of phase." In this case, the two sources of light literally cancel each other out. The result, then, is a series of bright and dark fringes across the screen. This is called an interference pattern, because the two sources of light are "interfering" with each other. Some detectors never register any light, while others register bright light all the time.

WAVE-PARTICLE DUALITY

Light behaves as an electromagnetic wave, but also as a stream of photons. Here, a screen is illuminated evenly by light from a light bulb. Each point on the screen receives the same intensity of light—or has an equal chance of receiving any photon.

1

(1) As a wave, the light's energy is delivered evenly across the screen

2

(2) Each photon delivers all its energy to one spot on the screen—but has an equal chance of landing anywhere

3

(3) Two waves interfere on the screen, producing dark fringes where they cancel out and light fringes where they reinforce each other

4

(4) The interference pattern still appears, built up from many individual photons that somehow interfere with themselves

This experiment has been known since 1800 (although without the detectors back then, of course) and was one of the main pieces of evidence in favor of the idea that light is a wave motion. The scenario becomes challenging when you repeat the experiment one photon at a time, using a single-photon light source, as before. In that case, only one detector will receive each photon, as was the case without the slits. But now, as the photons pass one by one from source to screen, through the slits, some detectors never receive any photons. Bizarrely, over time, the detectors' accumulated responses produce the same interference pattern overall, as you see when you shine a bright light. The inescapable conclusions seem to be that each photon is somehow traveling through both slits and "interfering with itself," and that photons really are behaving as waves and particles at the same time.

Wave-particle duality works both ways. Not only does light, traditionally thought of as waves, behave as a stream of particles, but also electrons, traditionally thought of as particles, behave as waves—as does any particle at the atomic scale. Carry out the double slit experiment with electrons instead of light and you will obtain the same results. Once again, with photons or electrons, the wave nature of these particles manifests as a probability that they will be found at particular locations. The probability is determined by the experimental setup (in this case, the arrangement of the two slits).

THE WAVE FUNCTION

To a quantum physicist, the waves in the experiments described above are not really light waves, but are "wave functions," mathematical expressions of probability. They have a value at each point in space and time, and that value determines the probability of finding a particle at that location at that moment. The wave function is a solution to the Schrödinger equation (see page 34), arguably the most important equation of quantum

$$i\hbar\frac{\partial}{\partial t}\Psi(\mathrm{r},t)=\hat{H}\Psi(\mathrm{r},t)$$

mechanics. The Schrödinger equation can be thought of as equivalent to Newton's Laws of Motion at the atomic scale.

Wave functions behave in the same way as any other waves. They can move through space and time, they can be reflected, they can spread out, and the undulations of several or many of them can add up—just as so many water waves can pass across a pond simultaneously, their disturbances adding up and resulting in choppy water. In quantum mechanics, one particular type of wave is very important: the standing wave.

The Schrödinger equation, containing the wave function, Ψ (the Greek letter "psi"), does not look too complicated. But the term $\hat{H}$, called the Hamiltonian operator, varies from situation to situation and can be especially complicated. The interpretation of quantum mechanics that is currently most widely accepted is that entities exist as a superposition (mixture) of all possible solutions of the wave function until they are observed or in some other way interact. At this point, the wave function "collapses" and the state particle, one of the possibilities, is "decided." In the two-slit experiment, then, the light really does travel as a wave (of probability) until it hits the screen. It delivers all its energy to one point, when and where the wave interacts with the screen, behaving as a discrete object: a particle.

The familiar conception of a wave is of a freely moving undulation, like a water wave moving across a pond or a sound wave speeding through the air. But in many situations, waves are "confined" and constrained in some way. For example, the ends of a guitar string are unable to move; when you pluck the string, waves travel up and down the string and are reflected at each end (see box). The waves interfere (their undulations are superimposed) and the result is a distinct and enduring pattern: a standing wave. A standing wave has "nodes," where no undulation occurs (such as the tethered ends of a guitar string), and "antinodes," where there is maximum undulation. As a result of these restrictions, only certain patterns are "allowed," those that satisfy the constraints (the tethered ends, in the case of the guitar string).

There are many possible standing wave patterns in any situation—and generally, several or all of the patterns occur simultaneously. Typically, the simplest pattern will dominate and the other modes will be less prominent. For a guitar, the particular mixture of different modes of standing wave—in the strings and also in the instrument's body—is what gives the instrument its characteristic sound. Standing waves are commonplace in everyday life, and in musical instruments in particular; for example, they form in the air inside woodwind instruments and in the skins of drums. They are also commonplace in the quantum world, as the wave functions of particles confined in some way, such as an electron bound to a nucleus by electrical attraction.

Erwin Schrödinger applied his equation to the hydrogen atom, which has just one electron. The result, the electron's wave function, consists of a series of possible three-dimensional standing waves the electron could form around the nucleus. Just as with the guitar string, only certain patterns are allowed, which ties in nicely with the fact that electron energies are quantized. Schrödinger suggested the waves represented the density of electric charge, as if the electron were physically spread out across the wave. It was German physicist Max Born (1882–1970) who first suggested that the wave function is a function of "probability density," in 1926.

STANDING WAVES ON A GUITAR STRING

A guitar string has many possible standing wave patterns, created by waves traveling up and down the string and reflecting at each end. In every one, there has to be a node at each end, because the ends are fixed. The simplest pattern has one antinode in the center of the string; this mode is called the fundamental. The other modes are called harmonics. The next simplest standing wave, called the first harmonic, has one node in the center of the string, and therefore two antinodes. The next harmonic pattern has two nodes in addition to the nodes at the end, and three antinodes, and so on.

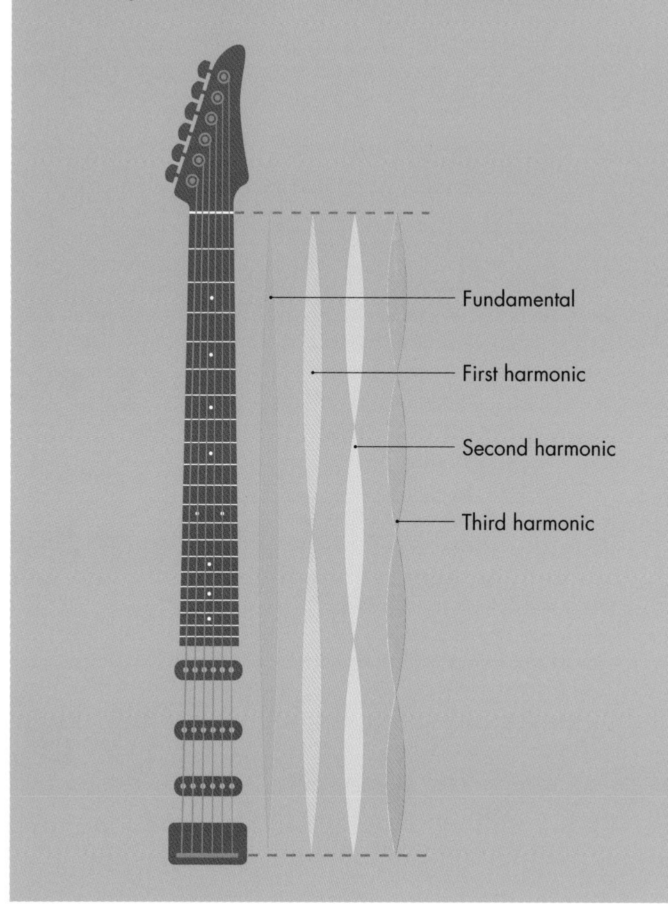

ELECTRON ORBITALS

The standing wave solutions of the Schrödinger equation, when considered as wave functions, determine the probability of finding the electron at any location. They define regions of space in which the electron will probably be found, and these regions are called "orbitals." The solution corresponding to the lowest electron energy ($n = 1$, the ground state) has a spherical shape. The probability of finding the electron varies with distance from the nucleus; it is zero at the nucleus, reaches a maximum at a particular distance, and then dies away. Note that the probability does not suddenly become zero—orbitals do not have definite edges—and that is why an atom does not have a definite radius.

The spherical form described above is called an s-orbital. At the next lowest energy ($n = 2$) there is another, larger, s-orbital. Once again, the probability varies continuously. In this case, the maximum probability lies at a greater distance than the maximum of the first s-orbital. At this second energy level, there are also three dumbbell-shaped "p-orbitals," each at right angles to the other two. In higher-energy levels, other, differently-shaped orbitals, called d- and f-orbitals, appear (see box).

Hydrogen has only one electron, to match the single proton in its nucleus. Add another electron (and another proton in the nucleus), and you have an atom of helium.

ATOMIC ORBITALS

The lowest-energy s-orbital (which corresponds to the ground state) is called the 1s orbital. The next available solution to the Schrödinger equation is the 2s orbital. At that second energy level, *n=2*, another kind of orbital appears. The p-orbital is shaped like a dumbbell, and there are three of them at each energy level at or above *n=2*. In the third level, *n=3*, yet another type of orbital makes its first appearance: the d-orbital. There are five d-orbitals at each energy level at or above *n=3*. At *n=4*, yet another type of orbital becomes available for occupation: the f-orbital. There are seven distinct f-orbitals at each energy level at or above *n=4*.

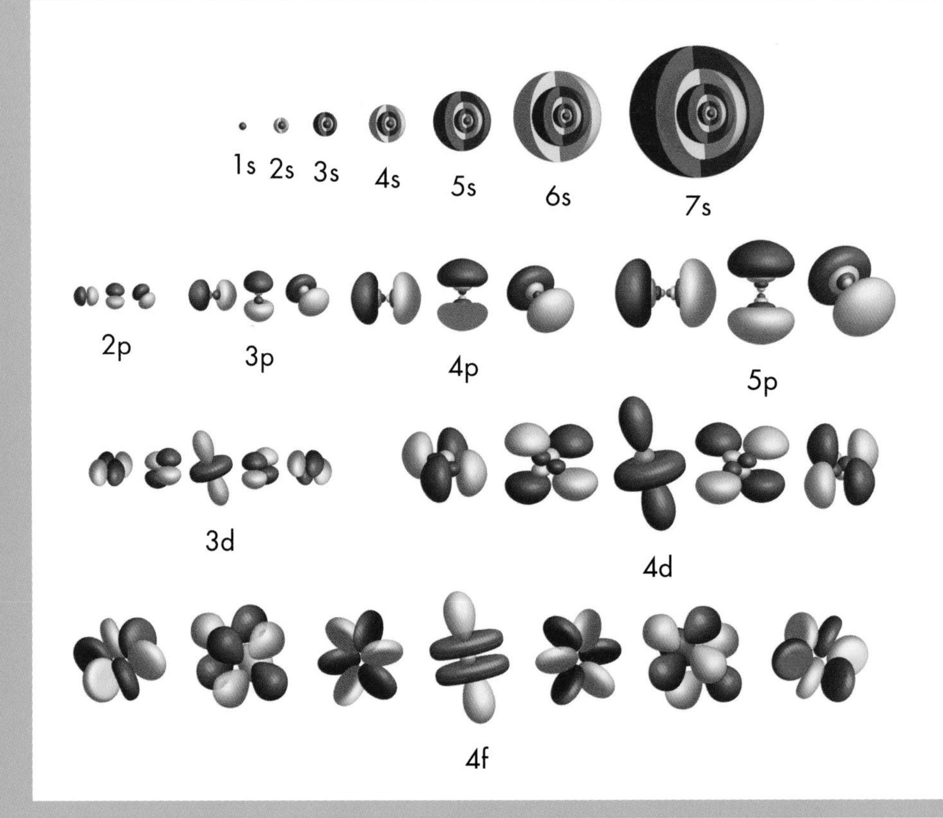

SPLIT LEVELS

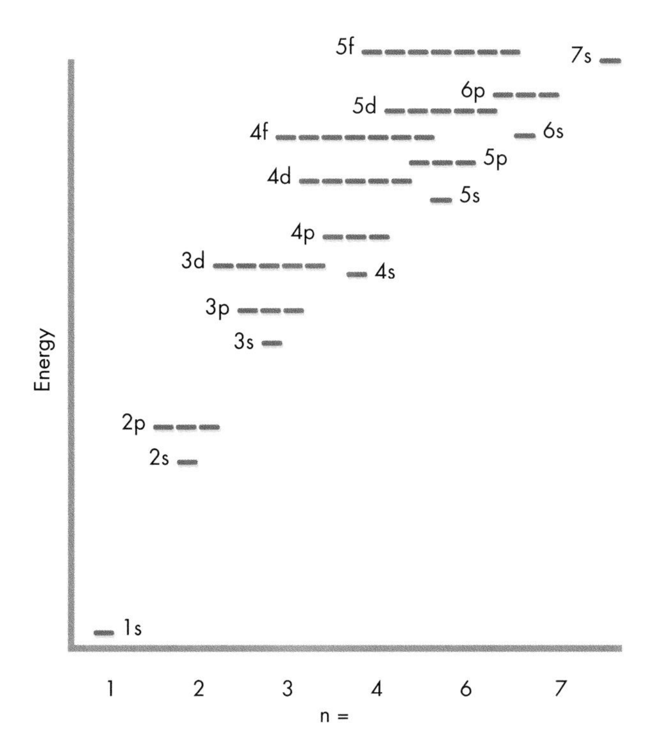

A graph of the available energy levels of a hydrogen atom. The numbers represent the shells, and the letters are the subshells (orbitals). Note how orbitals in the same shell have different energies—and how some energies in the fourth shell and above are lower than those in the shell below them.

Only up to two electrons are allowed in each orbital (see box, page 149), so a helium atom has a full 1s orbital; its "electron configuration" is written as $1s^2$. Add another electron and another proton, and you get the next element, lithium. Since the 1s orbital has its full complement of two electrons, the remaining electron occupies the 2s orbital—so its electron configuration is $1s^2\,2s^1$. As you add protons into the nucleus and corresponding numbers of electrons around it, the orbitals fill up to increasing energy levels. So, for example, boron has five electrons: two in the 1s-orbital, two in the 2s-orbital, and one in a 2p orbital. Electrons in the 2p orbital have a little more energy than those in the 2s orbital, despite the fact that they are both at the same ($n = 2$) energy level. This is because of interactions between the electrons, which cause the energy level to "split." So at each level, the s-orbitals are filled first, then the p-orbitals, and then (if present) the d-orbitals and the f-orbitals.

Atomic physicists and chemists refer to each energy level as a "shell" and the orbitals as "subshells." The number and arrangement of electrons in the outermost shell determine how an atom interacts with other atoms, and so give rise to the chemical properties of the elements. For example, an atom with a full outer shell will be stable and will not interact easily with other atoms. A helium atom has a full outer shell (its only shell), because it has the maximum two electrons in its 1s-orbital ($1s^2$). Helium is an unreactive element for just this reason. Similarly, a neon atom has a full outer shell, at the second energy level (two electrons in each of the 2s- and 2p-orbitals); it, too, is very unreactive. The ways in which elements' chemical properties depend upon the electron configurations of their atoms' outer shells will be explored in detail in chapter three.

Despite the weird and wonderful shapes of the orbitals, an isolated atom will always be a sphere. The three p-orbitals, for example, are "spherically symmetrical" overall, because they are identical and at right angles to each other. Even if there is only one electron present in all three, it has an equal chance of being in each of them, so the overall "probability cloud" is still a sphere. The same goes for the d- and f-orbitals: they, too, are spherically symmetrical overall. However, the shapes of orbitals often do become apparent. When atoms join together to form molecules, they do so by sharing electrons in "molecular orbitals." These merged orbitals determine the shapes of the molecules. Molecular orbitals will be explored in chapter four.

THE ATOMIC NUCLEUS

The two types of particle that the atomic nucleus is made of are both much more massive than the electrons, high above in their orbitals. In most nuclei, those protons and neutrons are tightly bound, making the nucleus stable. But some nuclei are unstable, giving rise to the phenomenon of radioactivity, in which a nucleus changes, or "decays." Because the nucleus is subject to the probabilistic laws of quantum mechanics, it is impossible to predict when a particular nucleus will decay.

The total number of protons and neutrons making up the nucleus is called the atomic mass number, or just the nucleon number. Each unique combination of protons and neutrons is called a nuclide. The number of protons in the nucleus is called the atomic number, and it defines the chemical element to which the atom belongs; each element has a unique atomic number. All oxygen atoms have eight protons, for example, so the atomic number of oxygen is 8.

Although the number of protons is the same for all atoms of an element, the number of neutrons is not. Different nuclides with the same number of protons but different numbers of neutrons are called isotopes. Because they have the same number of protons, a set of isotopes all belong to the same element, for example, oxygen (see page 57).

Every element has several isotopes, so there are many more nuclides than elements. In fact, there are only about 90 naturally occurring elements, but there are more than 330 naturally occurring nuclides. About 250 of those nuclides are stable, but the remaining 80 or so are not.

NUCLEAR INSTABILITY

There are two powerful forces at play in the nucleus. In many nuclides (the 250 or so stable ones), the forces are in balance. However, in others they are out of balance, and that can put the stability of the nucleus at risk. This can cause the nucleus to change, to attain lower-energy, and often more stable, states. One of the two forces governing the nucleus is the electrostatic force, the force between electrically-charged particles. Under the influence of the electrostatic force, two particles carrying the same kind of charge (a "+" with a "+" or a "−" with a "−") will repel, while two particles carrying different kinds of charge ("+" with "−" or "−" with "+") will attract each other. The closer two charged particles are, the greater is the repulsion or attraction between them. In the nucleus, the protons all carry positive charge, and they are extremely close together, so they repel strongly.

However, as noted in chapter one, the other force is a particularly strong attractive force that acts between all nucleons. It pulls together protons and protons, neutrons and neutrons, and protons and neutrons. This attractive force is stronger than the electrostatic repulsion, and it wins out, holding the nucleus together. But the nuclear force has an extremely limited range. As a result, above a certain diameter of nucleus, protons at extreme sides will be pushed apart more strongly than they are pulled together. The result is that larger, heavier nuclei are generally less stable than small ones. Neutrons are key to the stability of the nucleus, especially as the size of the nucleus grows. They contribute to the attractive nuclear force but do not contribute to the repulsive electrical force (because they are neutral, uncharged). They act as a kind of glue, helping to hold the nucleus together. If you add protons to a nucleus without adding enough neutrons, you put the forces out of balance, and that can make the nucleus unstable.

There are several different ways in which a nucleus can shift to a lower-energy, potentially more stable state. In each case, the nucleus emits, or radiates, energy and sometimes particles. Collectively, these processes are called radioactivity.

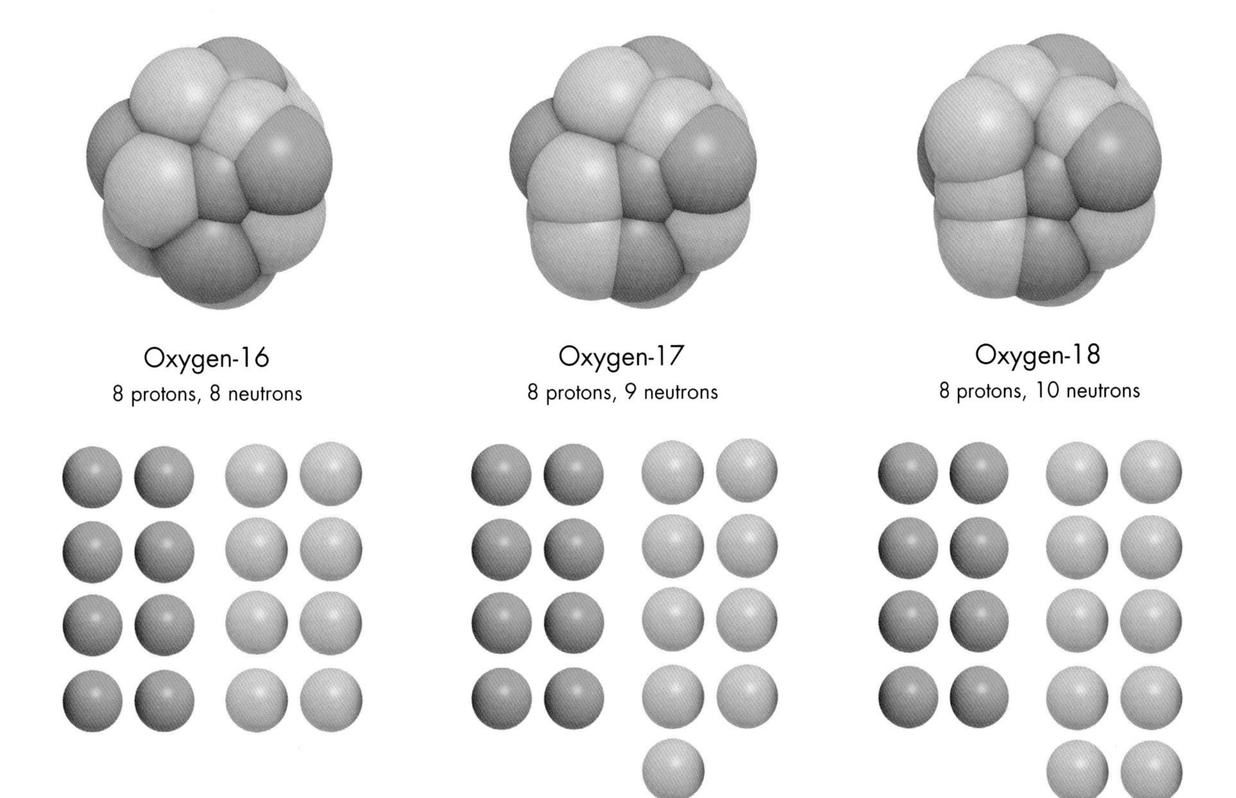

There are three isotopes of oxygen, the most common one being oxygen-16, whose atoms have eight neutrons plus the eight protons, giving them a mass number of 16. About one in every 500 oxygen atoms is oxygen-18, which has ten neutrons—and about one in every 2,500 oxygen atoms is oxygen-17, with nine neutrons.

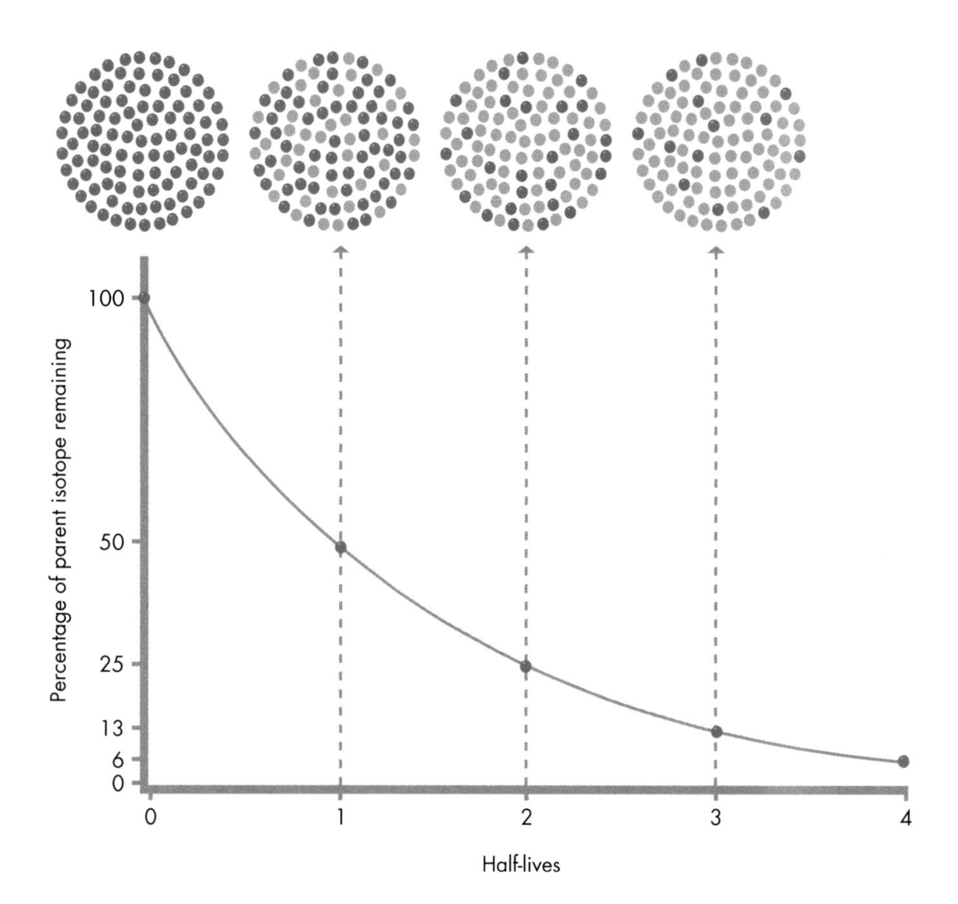

Each atomic nucleus in a radioactive element has the same chance of decaying at any time. As a result, the time it takes for half the nuclei to decay is always the same, however many you begin with. For the isotope uranium-238, this time, called the half-life, is about 4.5 billion years. After 9 billion years, two half-lives, only one-quarter of the original nuclei remain intact.

RADIOACTIVITY

Uranium atoms have the heaviest nucleus of any element commonly found on Earth. Each uranium nucleus has ninety-two protons. The most stable isotope of uranium is uranium-238, which has 146 neutrons (238 = 92 + 146). It is the most stable uranium isotope, but it is not completely stable. If you collected together a hundred uranium-238 atoms and waited 4.5 billion years, fifty of them would have decayed. After another 4.5 billion years, only twenty-five of the original hundred would still be undecayed. So 4.5 billion years is the half-life of uranium-238.

A uranium-238 nucleus decays by throwing out an alpha particle, a tightly bound unit of two protons and two neutrons. As a result of this "alpha decay," the nucleus loses two protons, and so the atom is no longer an atom of uranium. Instead, it is an atom of thorium, the element that has an atomic number of 90. The new nuclide is thorium-234 (the atomic number has reduced by two to 90, and the mass number has reduced by four to 234). The alpha particle released from the uranium nucleus shoots out at high speed and will probably be absorbed by another nucleus. It may have a very different fate, however, and find a separate existence by grabbing two electrons from another atom (two to match the number of protons). If this happens, it forms an independent, much smaller atom, an atom of the element helium. In fact, the source of nearly all the world's helium—to fill party balloons and MRI scanners—is alpha particles produced by radioactive decay underground, which have become helium atoms by grabbing hold of electrons.

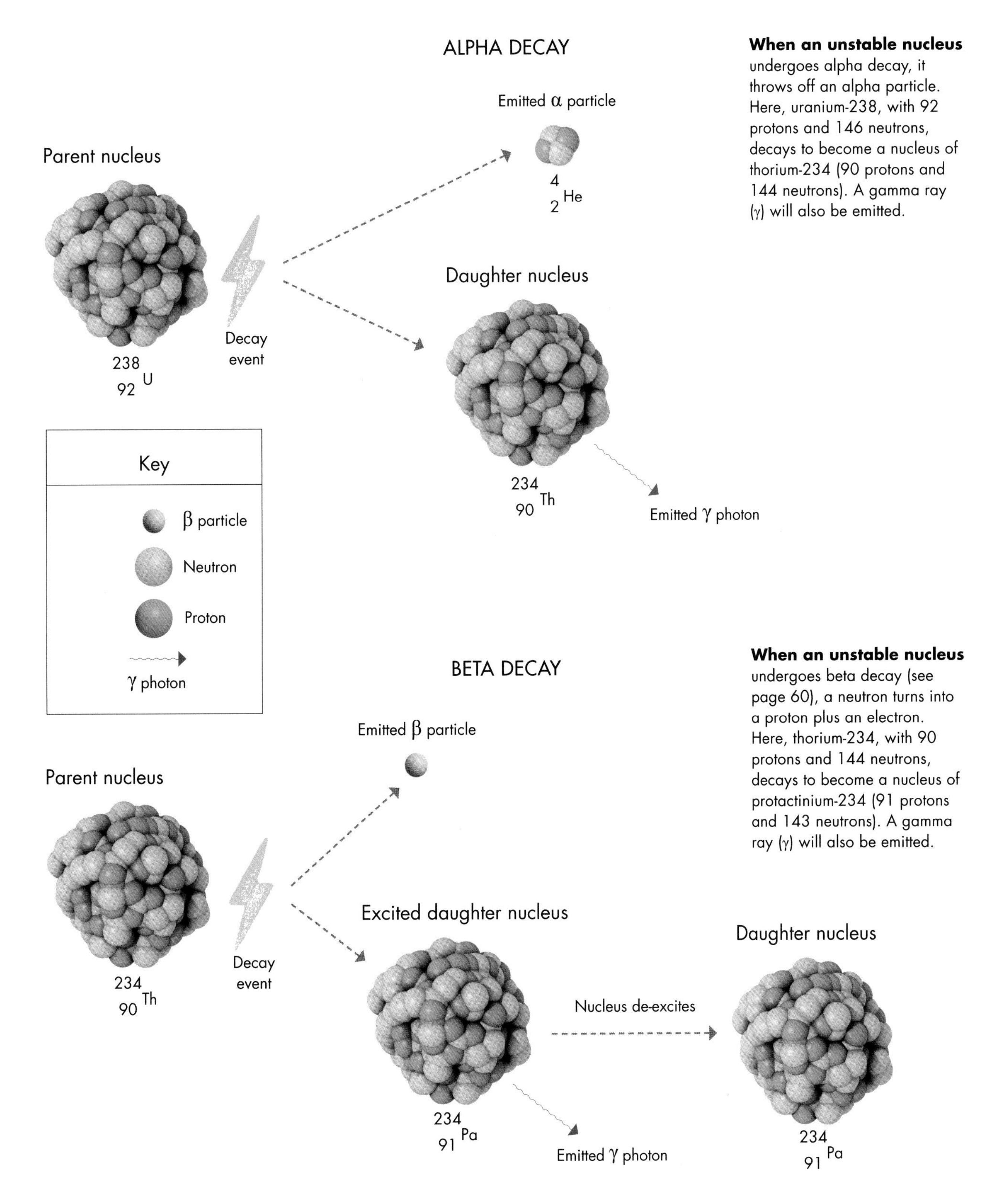

When an unstable nucleus undergoes alpha decay, it throws off an alpha particle. Here, uranium-238, with 92 protons and 146 neutrons, decays to become a nucleus of thorium-234 (90 protons and 144 neutrons). A gamma ray (γ) will also be emitted.

When an unstable nucleus undergoes beta decay (see page 60), a neutron turns into a proton plus an electron. Here, thorium-234, with 90 protons and 144 neutrons, decays to become a nucleus of protactinium-234 (91 protons and 143 neutrons). A gamma ray (γ) will also be emitted.

The Geiger-Müller tube is the most commonly used detector of radioactivity. The alpha particles (each one a clump of two protons and two neutrons), beta particles (fast electrons created in the nucleus), and gamma rays produced by radioactive decay can knock electrons off nearby atoms, creating an electrically-charged atom: an ion. If air becomes ionized, it can conduct electricity, and detectors of radioactivity work by detecting an electric current in ionized air. Although only half of the uranium atoms present will decay over 4.5 billion years, there are trillions and trillions of them in this sample, so many decay every second.

The "daughter nuclide" produced by the decay of the uranium-238 nucleus, thorium-234, has less energy than its parent, but it is actually far less stable. It has a good chance of decaying in a matter of weeks. This newly slimmed down but unstable nucleus will undergo a different decay process, called beta decay. In this case, a neutron in the nucleus actually changes its identity to become a proton, producing an electron in the process. The new electron, called a beta particle, shoots out of the nucleus at high speed into the world at large, passing its more sedate kin in their orbitals.

Note that there is, overall, no change in the amount of electric charge during beta decay. The neutron has no overall electric charge, and while the newly-created proton is positively charged, the newly-created electron is negatively charged.

What is known as the "conservation of charge" is always observed when subatomic particles interact. Note, too, that with one more proton, this new daughter nuclide now has an atomic number of 91, so the atom is no longer an atom of thorium. The new nuclide is protactinium-234. Once again, note that although the atomic number has increased by one, to 91, the mass number remains the same at 234, because the mass of the new proton is almost identical to the mass of the old neutron.

In both alpha and beta decay, the nucleus ends up in a lower-energy state—otherwise the decay would not have taken place. Some of the energy is needed to release an alpha particle or to produce an electron (beta particle), but the remaining energy is liberated as electromagnetic radiation. This is similar to the way in which electrons produce light when they fall from a higher to a lower energy level (see page 68). But the quantities of energy involved are much higher in the nucleus than in the electron orbitals, because the nucleus is a much more intense environment. As a result, the electromagnetic waves have very high frequencies, and each photon carries much more energy than the visible light photons produced by electrons. This high-frequency, high-energy radiation is called gamma radiation. In some cases, a nucleus will simply shift from an "excited," higher-energy state to a more stable, lower-energy state, producing gamma radiation without producing alpha or beta particles. Indeed, the daughter nuclide left behind after thorium-234 undergoes beta decay—protactinium-234—is in an excited or "metastable" state (protactinium-234m). It may decay to a lower-energy state by emitting gamma radiation, although most protactinium-234m nuclei will undergo beta decay, resulting in uranium-234.

RADIOACTIVE EMISSIONS

The decay of an unstable nucleus produces three types of radiation: alpha, beta, and gamma. The heavier alpha particles are more likely to be absorbed by other nuclei, so they are the least "penetrating." They can be stopped by a sheet of paper. Beta particles are faster and lighter; they will penetrate through paper but can be stopped by a thin sheet of metal—aluminum, for example. Gamma rays are a powerful form of electromagnetic radiation, and they can only be stopped by a thick layer of lead or other dense materials. The dangers and benefits of radioactivity are explored in chapter six.

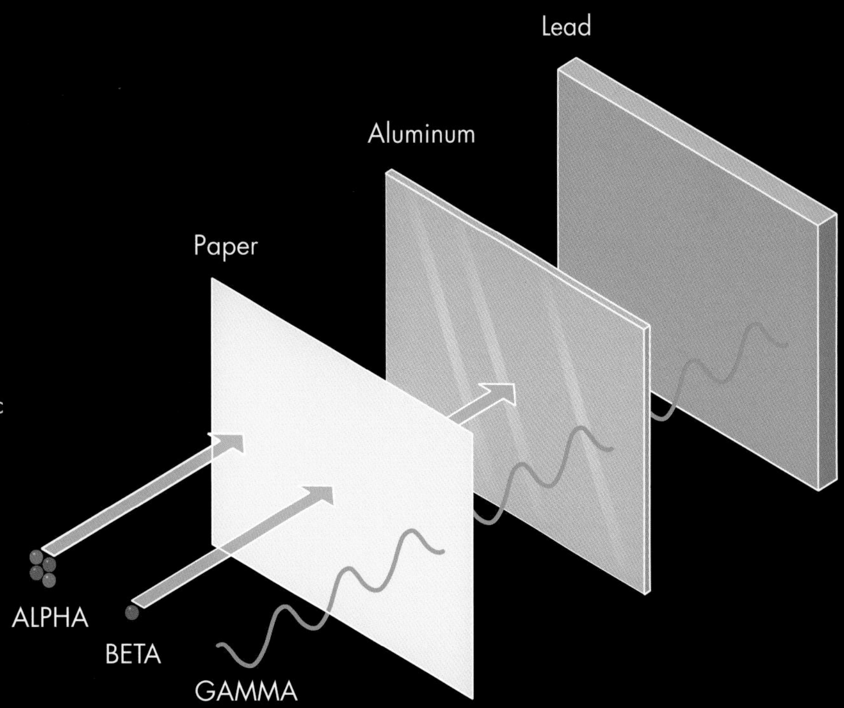

THE QUANTUM NUCLEUS

It would be natural to wonder, if a uranium-238 nucleus is unstable, why has it got a 50 percent chance of remaining unchanged, undecayed, after 4.5 billion years? Similarly, why does an unstable thorium-234 nucleus typically take several weeks to decay? Why do these unstable nuclei not decay right away? The answers lie in quantum mechanics. For, just as an atom's electrons are subject to the strange laws of quantum physics, so are the protons and neutrons that make up the atomic nucleus. The nucleus has a wave function and well-defined energy levels, and its behavior is determined by probability. As a result, and without any influence over each other, exactly half the atoms in any sample of uranium-238 will decay after 4.5 billion years. Which nuclei will decay is purely a matter of chance.

In the famous Schrödinger's Cat paradox, a radioactive nucleus triggers the opening of a vial of deadly poison if it decays. Quantum states exist simultaneously until they are "observed," so the nucleus is both decayed and undecayed and the cat is dead and alive, until the box is open.

HALF-LIVES

All of the 330 or so naturally occurring nuclides are plotted here on a chart. The vertical axis represents the atomic number, the number of protons, while the horizontal axis represents the number of neutrons. Nuclides colored black are stable. The other colors represent the nuclides' half-lives, from blue (most stable, longer half-lives) to yellow (least stable, shorter half-lives. Notice how, as the size of the nucleus increases, more neutrons are required to "glue" the nucleus together and make it stable.

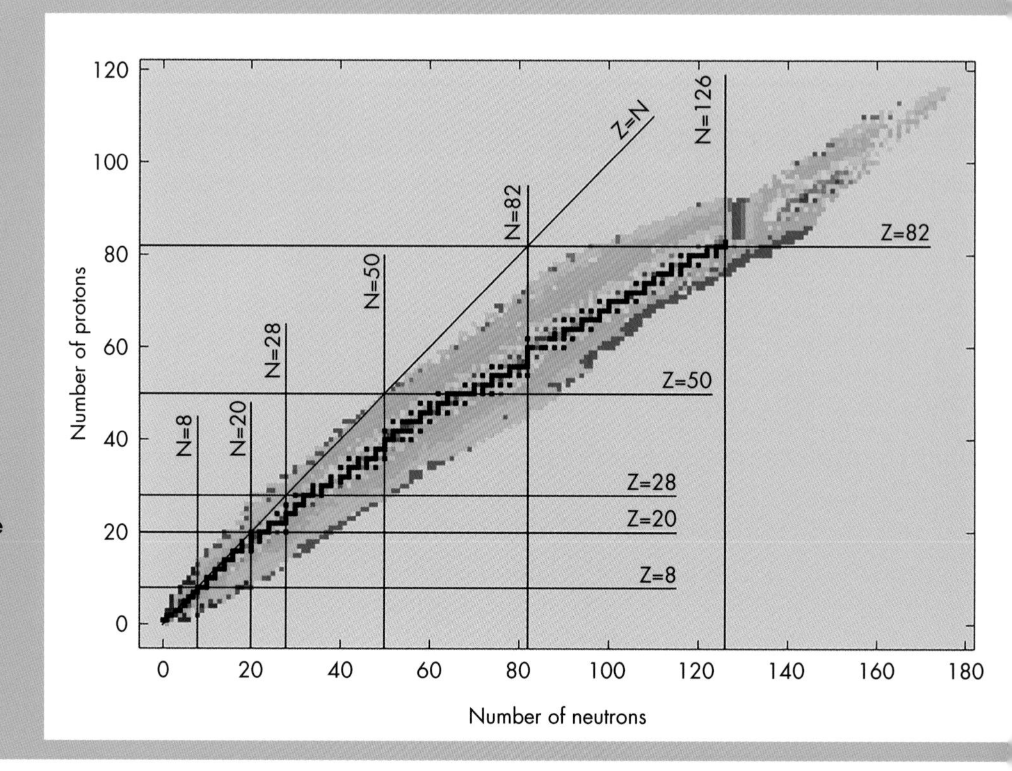

QUANTUM TUNNELING

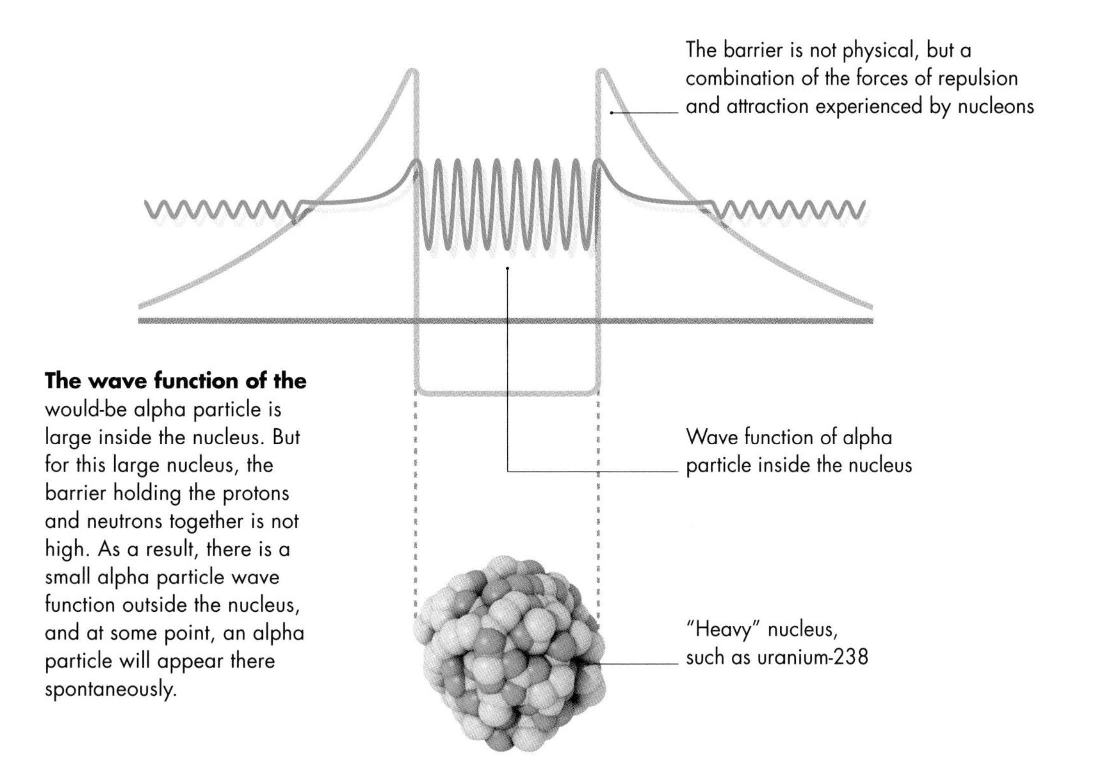

Alpha decay occurs as a result of a quantum mechanical phenomenon known as tunneling (see also page 123). The combination of the two forces holding the protons and neutrons in the nucleus creates a "barrier" that normally keeps the particles from leaving the nucleus. The height and width of the barrier is determined by the number of nucleons present and by the particular combination of protons and neutrons. The wave functions of all the particles present in the nucleus interact with the barrier and become standing waves bouncing around inside the boundary of the nucleus—in a similar way to how electrons form standing waves. The overall wave function of the nucleus is a superposition (combination) of the wave functions of all the protons and neutrons present and, indeed, of all the possible combinations of those particles. So, even though it may be as yet a nonexistent particle, there is a wave function for an alpha particle (or many alpha particles) inside the nucleus.

This wave function forms a standing wave as it reflects off the barrier. If the barrier was of infinite height and width, the wave function would stop abruptly at the barrier—in the same way as the tethered ends of a guitar string prevent any vibration there. However, with a finite barrier, the wave function exists inside and beyond the barrier, albeit much diminished. In other words, there is a small but real probability that the particle might be found outside the nucleus. The lower and narrower the barrier (the more unstable the nucleus), the greater the chance that an alpha particle will spontaneously tunnel through the barrier and appear on the other side.

With an understanding of atomic number, of isotopes, and of how electrons arrange themselves around the nucleus, it is possible to organize the possible nuclides into a neat chart—the periodic table. This is the subject of the next chapter.

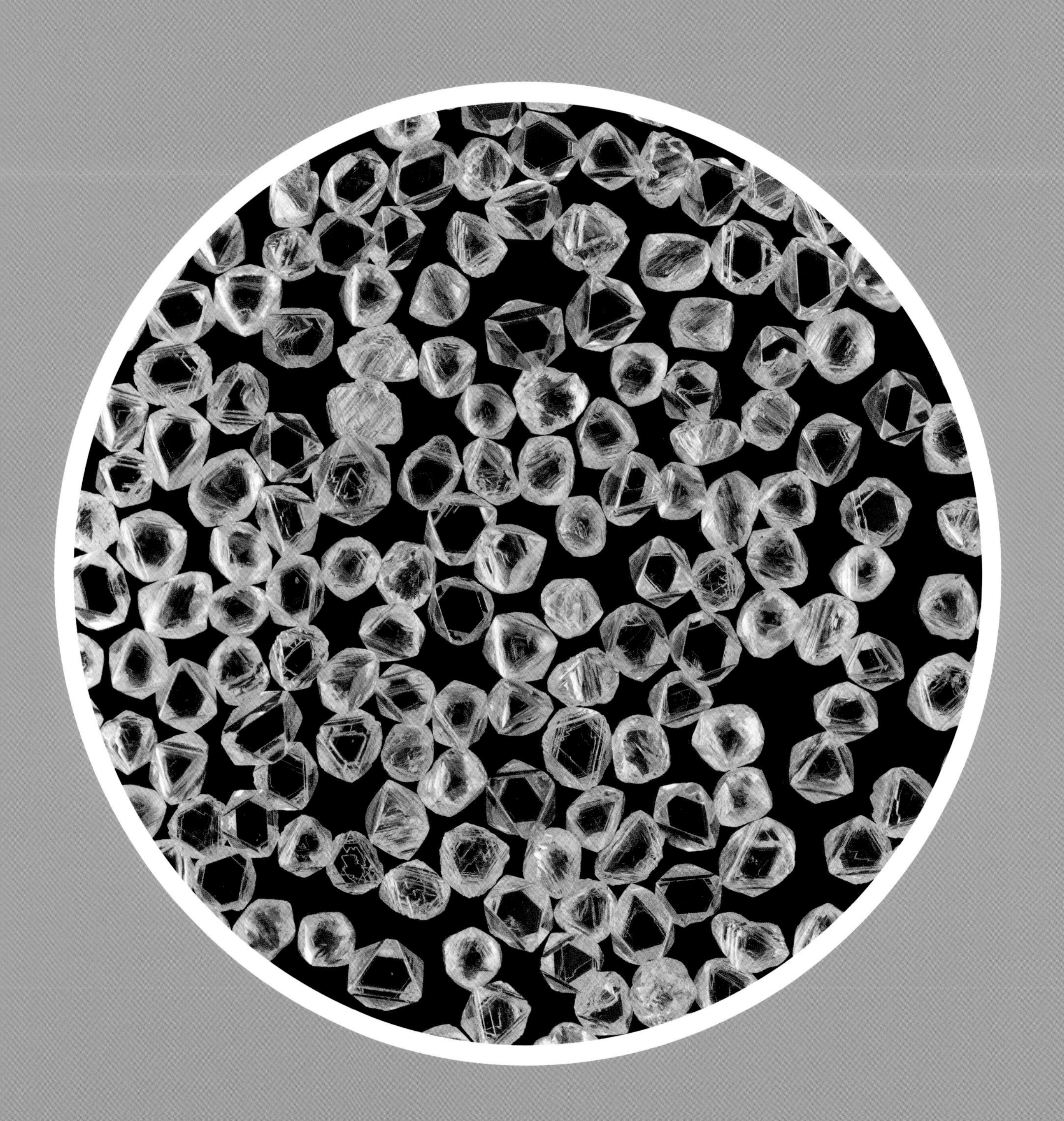

CHAPTER 3
ATOMIC IDENTITIES

Everything around you is made of atoms of ninety or so chemical elements. All the atoms of a particular element have the same number of protons in the nucleus (and that same number of electrons surrounding the nucleus). In other words, the proton number determines an atom's identity. The chemical properties of each element—how it interacts with other elements—are determined by the arrangement of the outermost electrons in their orbitals.

Rough-cut diamonds.
Pure diamond is a material made of only carbon atoms, held together in a rigid crystal structure. Most mined diamonds contain impurities—other types of atoms in the crystal structure—that result in slight discoloration.

IDENTIFYING ELEMENTS

Around ninety elements—ninety types of atom—are found naturally on Earth. Uranium (element 92) is the heaviest element for which there is a stable isotope, and there are two unstable elements lighter than uranium that are not found naturally. However, the exact number is vague, because tiny amounts of elements heavier than uranium are found in extreme circumstances. Whatever the number, it is certainly large. How can you tell one from another?

There is a great deal of variety in the properties of pure elements. For example, at room temperature some are invisible gases, while others are shiny, metallic solids or brightly colored liquids. Some elements are highly reactive, others inert; some have extremely high boiling points, others extremely low ones. The exact combination of physical and chemical properties—a result of the configuration of electrons around the nucleus, and the number of protons and neutrons in the nucleus—can identify a pure element. So, for example, if you have a sample of a chemically reactive, colorless gas whose boiling point is -297°F (-183°C), then you have oxygen.

Native elements are elements that sometimes naturally occur in a more-or-less pure state. Each of the small samples shown here is made from many trillions of atoms, nearly all of which are the same. Each one will also contain millions or billions of atoms of other elements.

Sulfur

Silver

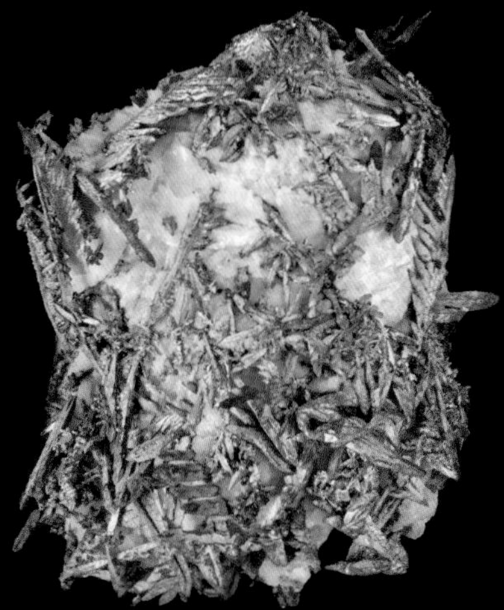

Most elements are rarely found pure. Instead, they exist in compounds, in which their atoms are bound tightly to atoms of other elements. Of the thirty or so elements that do sometimes exist naturally in their pure state, gold, copper, carbon, sulfur, and silver are relatively easy to identify by sight. To identify the majority of elements, which only exist in nature combined with other elements, you must first separate them into their pure state. Most metals, for example, exist as ores, their atoms typically bound to oxygen atoms. Smelting normally involves heating a metal with carbon, so that the carbon atoms can steal the oxygen atoms away (forming carbon dioxide molecules) and the pure metal is left behind.

Copper metal appears when the bonds between copper atoms and oxygen atoms in copper ore are broken in the presence of heat and carbon atoms in charcoal.

Graphite (carbon)

Native gold

Copper

SPECTROSCOPY

In many cases, heating a compound can cause it to dissociate into its elements, the atoms breaking away to form a vapor. One way of identifying the metallic elements present in a compound, the flame test, relies on this fact. The mystery compound is heated in a flame, releasing the metal atoms, which form a vapor. Electrons in the hot atoms are boosted up to higher-energy levels, and then fall down, emitting light of a characteristic frequency (and therefore color). The exact frequency of the light emitted depends upon the difference in energy between the two levels (see page 46)—and that is unique to each element.

To be sure that a particular element is present, scientists normally study the colored light in a spectroscope, which separates out the individual frequencies present (each one corresponding to a particular pair of energy levels). The same characteristic frequencies are behind many everyday phenomena, including the colors of fireworks and the orange color of sodium lamps used for some types of streetlights. Many of the elements discovered since the 1860s have been identified as new elements—or have had their status as "newly-discovered" verified—by variations of this technique, which is known as spectroscopy.

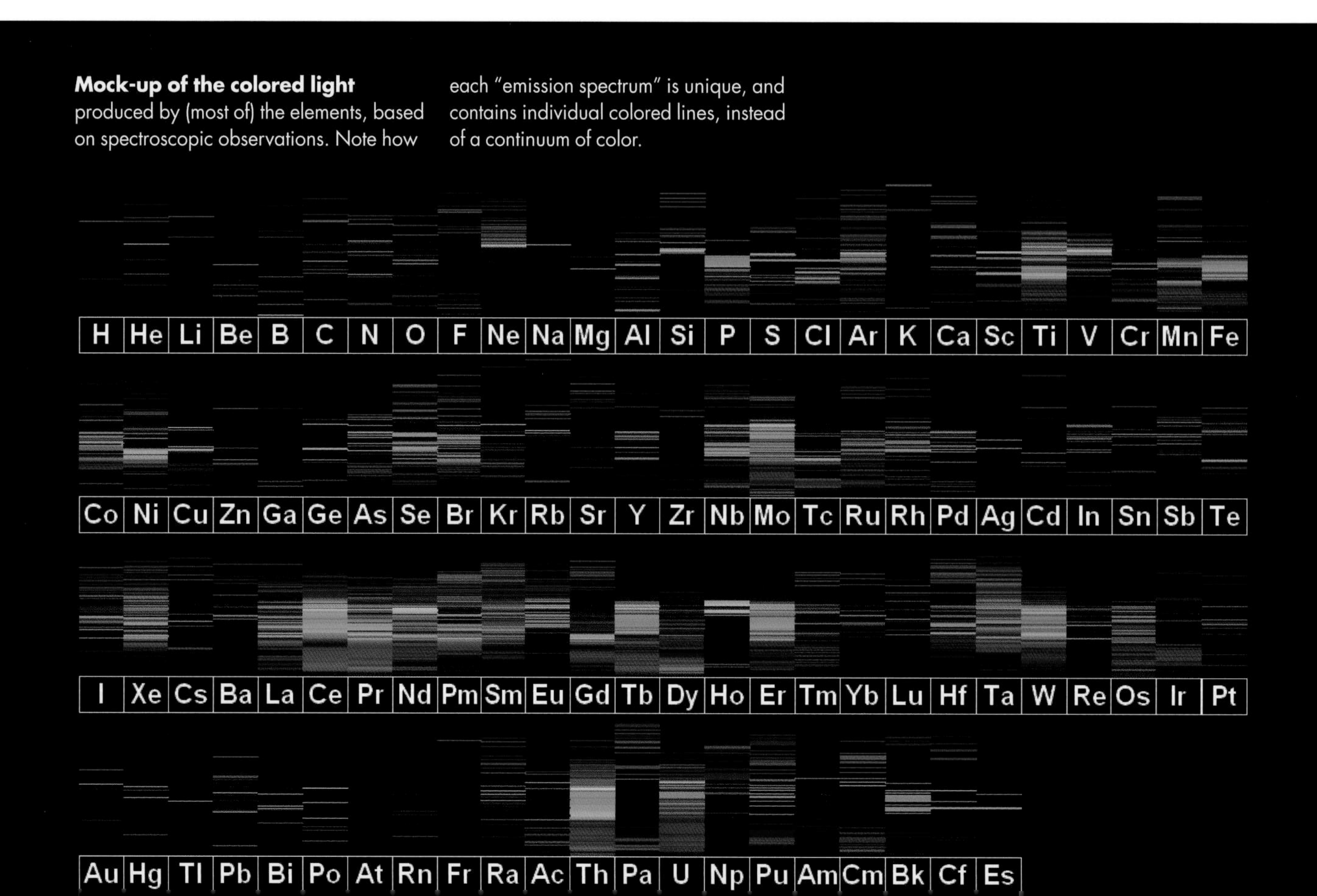

Mock-up of the colored light produced by (most of) the elements, based on spectroscopic observations. Note how each "emission spectrum" is unique, and contains individual colored lines, instead of a continuum of color.

SORTING BY MASS

Inside a mass spectrometer, atoms from a vaporized gas are ionized by an electron beam and then accelerated past a magnetic field, which bends their path. The lighter the ion, the greater the deflection.

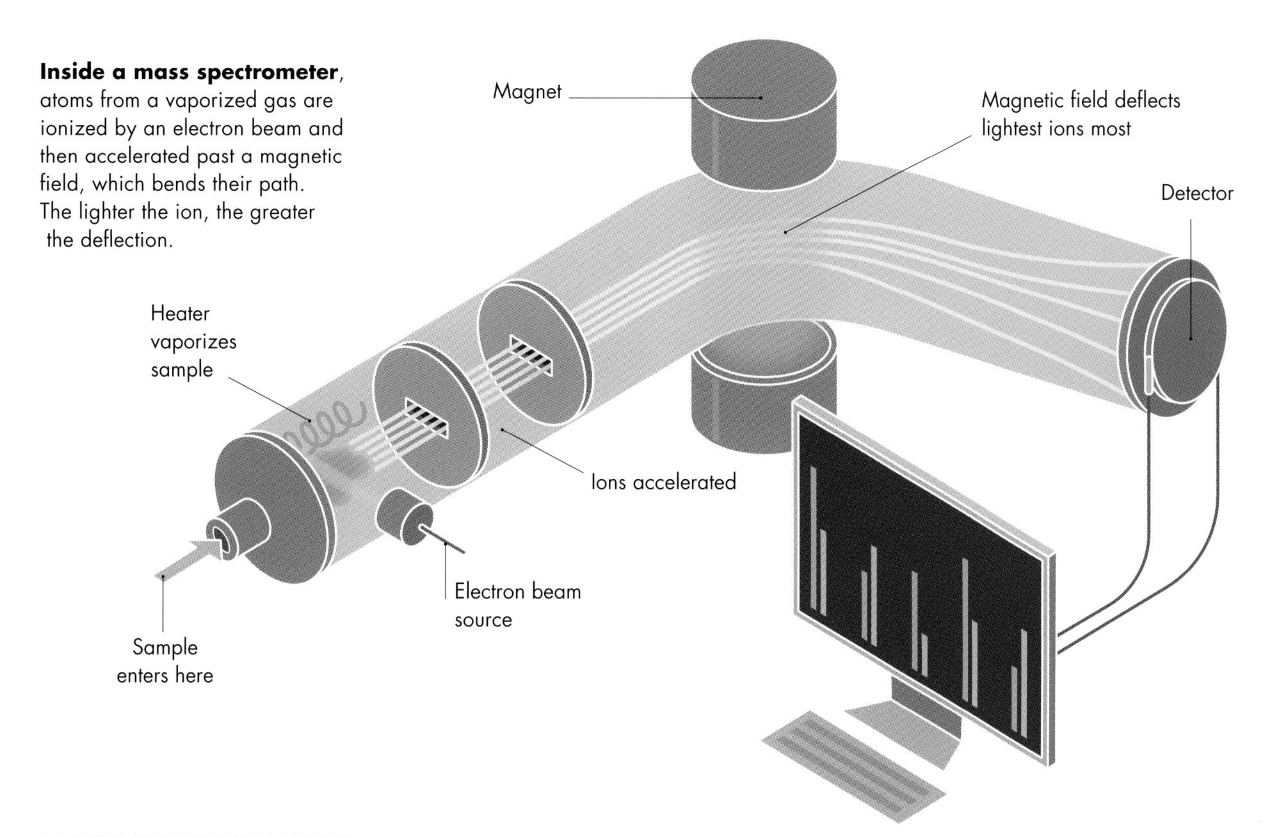

MASS SPECTROMETER

Another way of identifying elements normally combined in compounds is mass spectrometry. Inside an evacuated chamber—in other words, one from which the air has been removed—a sample for testing is first vaporized to break it into individual atoms. A high-powered electron beam knocks electrons off the atoms, turning them into positive ions. These ions are then accelerated by a strong electric field, and they pass along the chamber at speeds of several kilometers per second. Strong electric and magnetic fields inside the chamber force the ions to follow curved paths. Crucially, the heavier the ion, the less it will be deflected—just as blowing a passing tennis ball will change its course much less than doing the same to a passing ping-pong ball. A device at the far end of the curved chamber detects the ions as they arrive, and it is then possible to work out the mass of the ions passing along the chamber, and therefore which elements are present in the sample. Because the ions are separated according to their mass, this technique can even separate different isotopes (that is, different versions of the same element with the same number of protons but different numbers of neutrons). Mass spectrometry has many applications, including in forensics and in purifying a sample of uranium into its two main isotopes, only one of which is useful in nuclear power stations.

Identifying elements is a challenge that scientists have mastered using techniques such as those described above. A far greater challenge was to work out why there are elements at all—which scientists have now done. To answer this we must first find out where the elements came from.

THE ORIGINS OF THE ELEMENTS

All the matter around you is made of atomic nuclei plus electrons, often bound together as atoms (or ions or molecules). The number of protons in a nucleus determines the element to which its atom belongs. Some of the nuclei—and therefore some of the elements—date back to the first seconds and minutes after the beginning of time. Others were formed inside stars, and yet others in extremely energetic supernovas. The rest are the result of radioactive decay.

NUCLIDES FROM THE DAWN OF TIME

Atomic nuclei are composed of protons and neutrons, each unique combination referred to as a nuclide. Before they have electrons bound to them, nuclei are simply clumps of protons and neutrons—tiny objects that could be called "would-be nuclei." The first would-be nuclei were produced in the early stages of the universe. As far as we know—as far as the evidence and theories of modern cosmology tell us—the universe came from nothing, suddenly, 13.8 billion years ago. Quarks immediately "condensed" out of energy in extremely large numbers. After about a millionth of a second, most of those quarks had formed composite particles, each consisting of a triplet of quarks. These particles were protons and neutrons. An individual proton is the would-be nucleus of a hydrogen-1 atom—so, by default, hydrogen-1 was the first nuclide to come into existence.

Initially, there were equal numbers of protons and neutrons. However, free (unbound) neutrons decay to form a proton plus an electron, so there were soon many more protons than neutrons. In fact, within a few seconds, protons outnumbered neutrons by about seven to one. Over the next few minutes—within minutes of the beginning of time—many of the neutrons became bound to protons, forming nuclides heavier than hydrogen-1.

NEUTRON DECAY

A free (unbound) neutron is unstable. The decay of free neutrons resulted in a dramatic imbalance of protons and neutrons in the early universe. Note how there is no overall electric charge after the decay. For information on antineutrinos, see chapter seven.

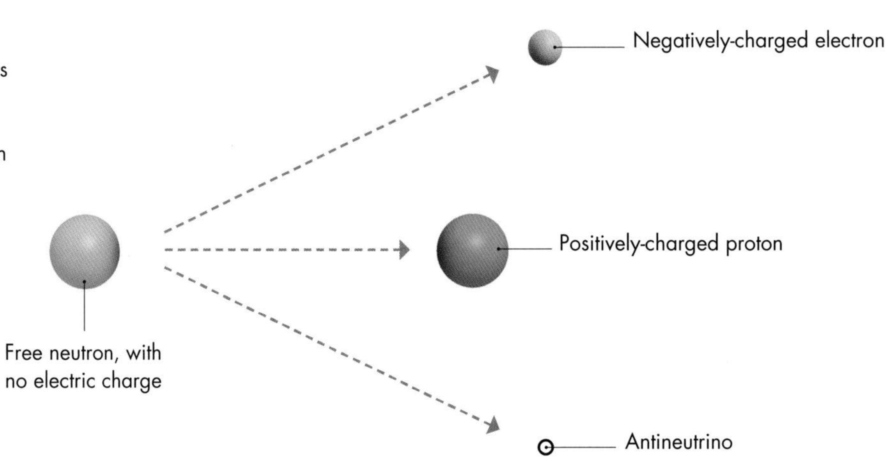

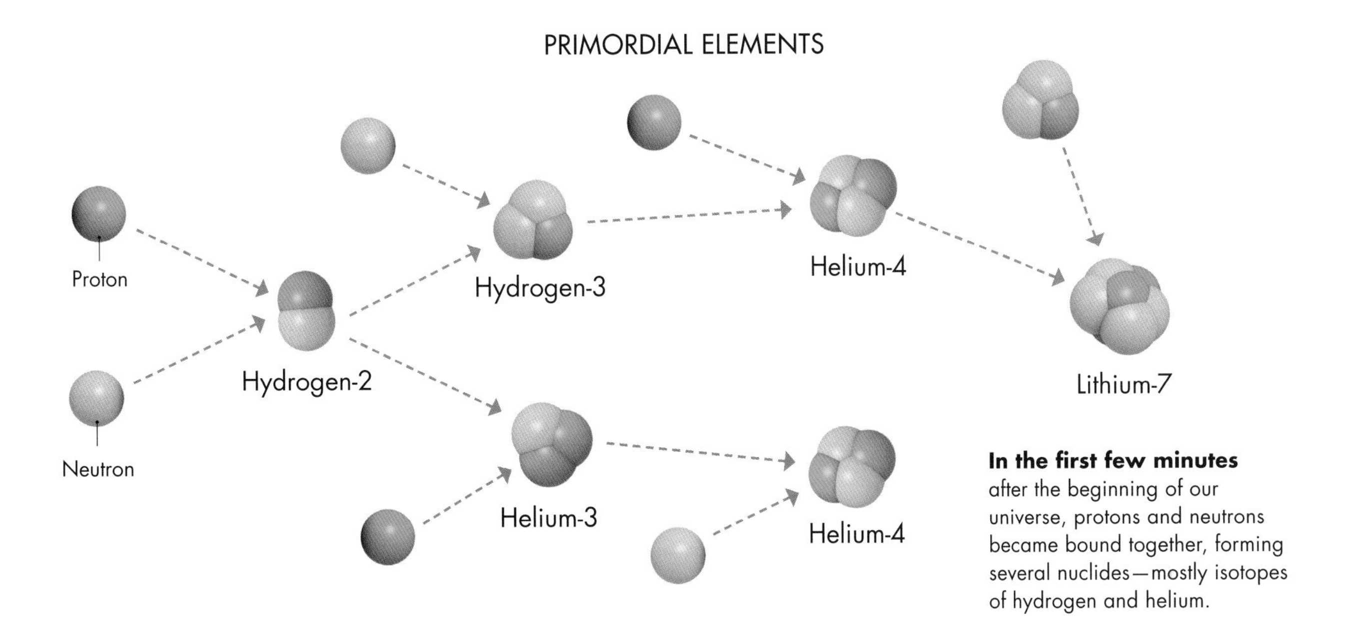

In the first few minutes after the beginning of our universe, protons and neutrons became bound together, forming several nuclides—mostly isotopes of hydrogen and helium.

One proton bound to one neutron makes hydrogen-2 (1p, 1n), which is also called deuterium. Add another neutron, and you get hydrogen-3 (1p, 2n), also called tritium. Add another proton instead, and you have helium-3 (2p, 1n). These were simply intermediaries for a much more stable nuclide, helium-4 (2p, 2n).

If the universe had expanded more slowly, all the neutrons would have ended up in would-be helium-4 nuclei, and the remaining protons would have been left behind as hydrogen-1—these would then have been the only two nuclides present. But the universe expanded extremely rapidly, so small amounts of deuterium and helium-3 also remained (the tritium was unstable and quickly decayed to form helium-3). In addition, a tiny proportion of the first would-be nuclei were lithium-7 (3p, 4n). Overall, though, hydrogen-1 and helium-4 accounted for around 99.9 percent of the would-be nuclei present after those first few minutes. And to this day, these two nuclides are by far the most abundant in the universe.

Electrons were also created in large numbers within the first millionth of a second. But in the intense frenzy of the early universe, both the high temperature and the radiation coursing around space were too great for the electrons to settle neatly into orbitals around the would-be nuclei, so no atoms could exist. Instead, the matter in the universe existed as "plasma," a mixture of negatively-charged electrons and positively-charged ions. A positively-charged ion is an atom with fewer electrons than protons. In this case, the atoms had no electrons at all; the plasma was a mixture of electrons and completely "naked" would-be nuclei. It was only after 380,000 years that conditions calmed down enough for the first atoms to appear. The would-be nuclei became atomic nuclei at last. With electrons bound to atoms, the universe became transparent—previously, any radiation was absorbed and then reradiated by free electrons, making space opaque and foggy. As well as being transparent, space was also dark, because the initially hot universe had cooled down, and there was nothing to produce light. That all changed about 200 million years later.

1

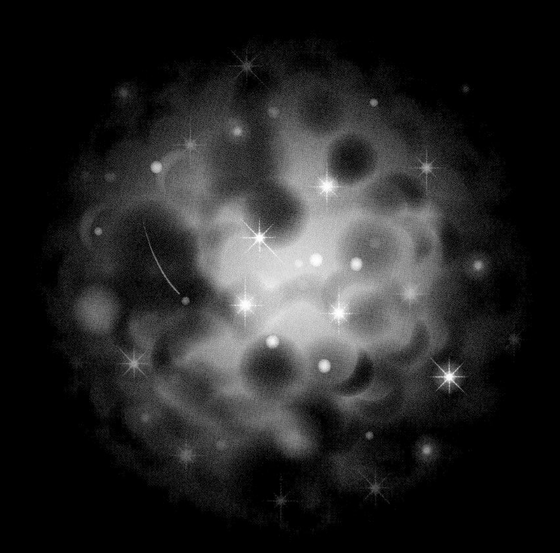

STAR BIRTH
The first generation of stars would have been much larger on average than later generations, but new stars being born in the universe right now form in the same way. Gravity pulls the densest regions of vast gas clouds into clumps (1). As the gravitational collapse proceeds, the gas heats up and the pressure at the center becomes so great that nuclear fusion begins. The heat causes the gas to expand, supporting the young "protostar" against further collapse. The energy released by the nuclear fusion heats the gas to a high temperature. The young star emits intense electromagnetic light and other radiation, and throws out a wind of charged particles into the space around it (2). These emanations clear the remaining gas around the star (3).

FIRST LIGHT

In the long dark ages of the universe, some regions of space were filled with huge clouds made almost entirely of hydrogen and helium gas. In other places, the universe was empty. Most of the hydrogen existed as molecules, each consisting of two atoms joined together (see chapter four), while the helium was made of individual atoms. The gases were extremely rarefied. There were a few thousand atoms or molecules per cubic centimeter, which makes it about as dense as the best vacuums scientists can produce here on Earth. Certain parts of the gas clouds were very slightly more dense than the average and, over millions of years, the mutual gravity between the atoms and molecules in these denser regions began to pull the hydrogen and helium mixture together.

This gravitational collapse resulted in spherical blobs of gas that grew ever more dense, and the energy released by the collapse heated up the gas at the centers of the blobs. The rising temperatures in the blobs of gas caused the electrons to leave their atoms, so the gas became plasma once again. Eventually, the temperature and the pressure inside these blobs of gas became high enough to force some of the tiny primordial nuclei to fuse together, creating new would-be nuclei. This "nuclear fusion" released enormous amounts of energy, heating the gas still further. This process had two effects. First, the hot gas expanded, buoying it up against further gravitational collapse. Second, the high temperature caused the enormous blobs to glow with incandescence, becoming fully-fledged stars—bright beacons in an otherwise dark universe.

This first generation of stars began with mostly hydrogen-1 (naked protons) and a little helium-4. For most of their lives, these stars simply made more helium-4 from the hydrogen-1 in a reaction called the proton-proton chain. This "hydrogen burning" is also the predominant reaction in most stars today, including our Sun. So the energy that sustains all life on Earth was released deep inside the Sun as the result of hydrogen nuclei being forced together to form helium.

2

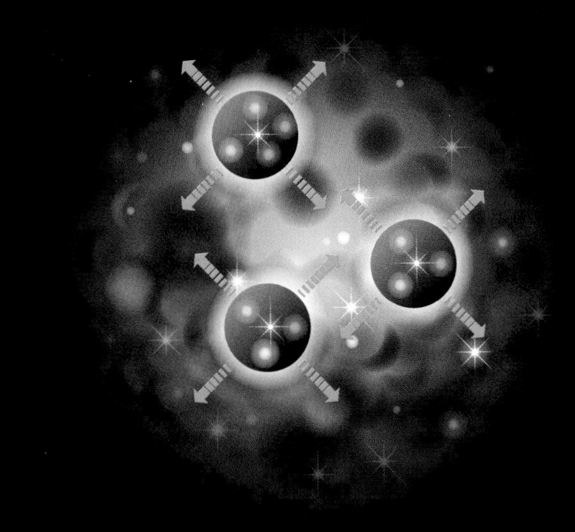

3

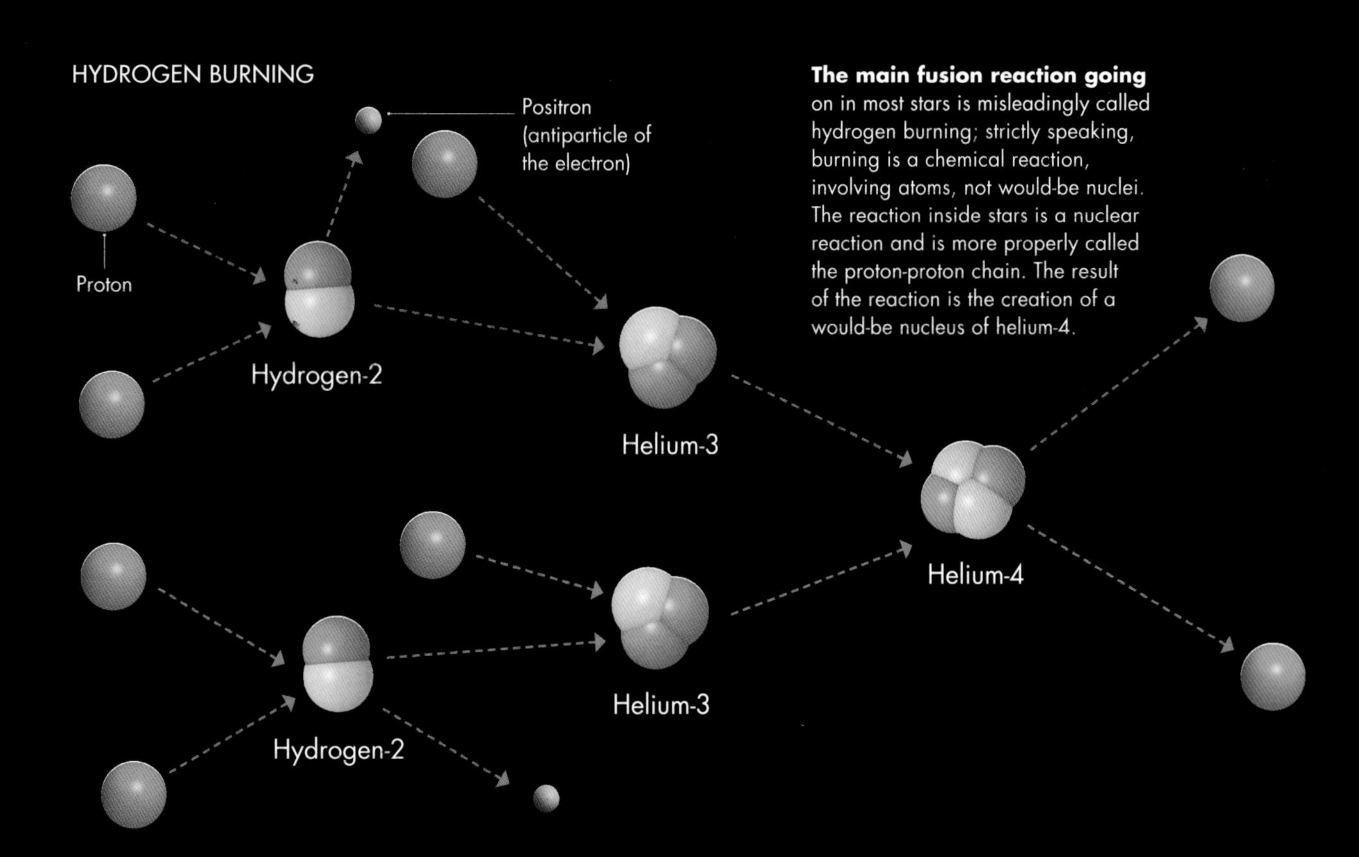

The main fusion reaction going on in most stars is misleadingly called hydrogen burning; strictly speaking, burning is a chemical reaction, involving atoms, not would-be nuclei. The reaction inside stars is a nuclear reaction and is more properly called the proton-proton chain. The result of the reaction is the creation of a would-be nucleus of helium-4.

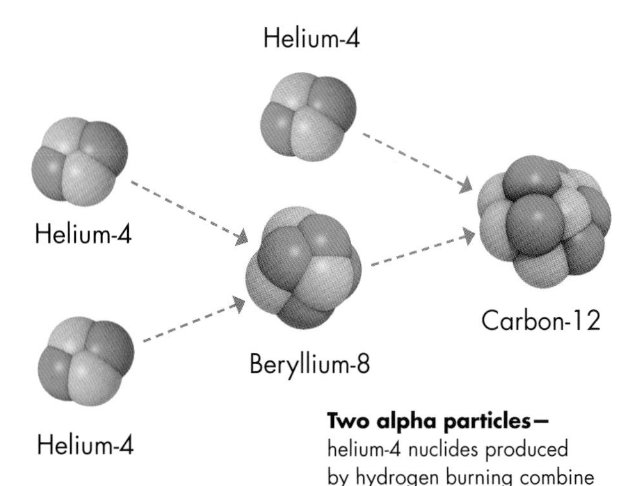

Two alpha particles— helium-4 nuclides produced by hydrogen burning combine to make beryllium-8; add another alpha particle and the result is one would-be nucleus of carbon-12.

Once most of the hydrogen in a star's core is used up—which will happen inside the Sun in a few billion years time—a new reaction can begin. This new reaction is known as the triple alpha process, in which three helium-4 nuclei combine to form carbon-12 (6p, 6n). The process gets its name from the fact that a naked helium-4 nucleus (2p, 2n) is identical to an alpha particle ("α," see page 58), and it takes three of those to make carbon-12. Other combinations of alpha particles beyond carbon-12 form different, heavier nuclides. All of them have atomic numbers that are multiples of two and nuclide numbers that are multiples of four; for example, oxygen-16 (8p, 8n; or 4α), neon-20 (10p, 10n; or 5α) and magnesium-24 (12p, 12n; or 6α). These only form in massive stars, where the temperature and pressure is enough to cause these reactions at the core.

Successive rounds of this alpha process take place, building heavier and heavier nuclides. In the meantime, another process creates yet more new nuclides. It is the result of free neutrons dashing around inside the star. When a free neutron hits a nucleus, it has a good chance of sticking to it, creating a nuclide of the same element (there has been no change to the number of protons) but with a mass number increased by one. So, for example, oxygen-16 (8p, 8n) would become oxygen-17 (8p, 9n). In some cases, the new nuclide would be unstable. It might undergo beta decay (see page 59), in which a neutron inside the nucleus decays to form a proton and releases an electron. If that happens, the nucleus's proton number increases, because there is one more proton. So in our example, oxygen-17 (8p, 9n) would become fluorine-17 (9p, 8n). There are not many free neutrons in a star, so this process is slow. It is called the s-process, and the "s" really does stand for "slow." The s-process occurs in the last few tens of thousands of years of a star's lifetime.

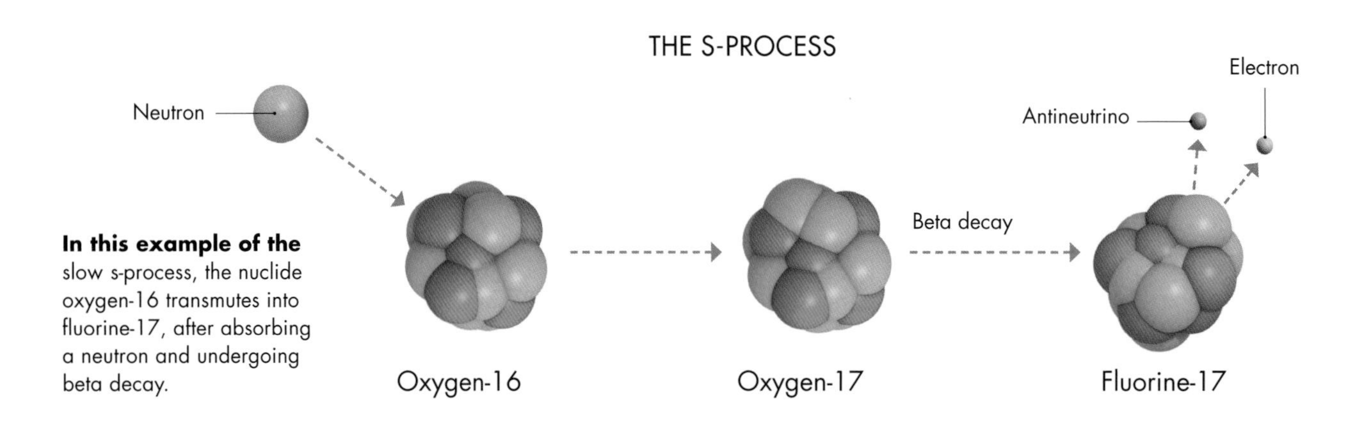

In this example of the slow s-process, the nuclide oxygen-16 transmutes into fluorine-17, after absorbing a neutron and undergoing beta decay.

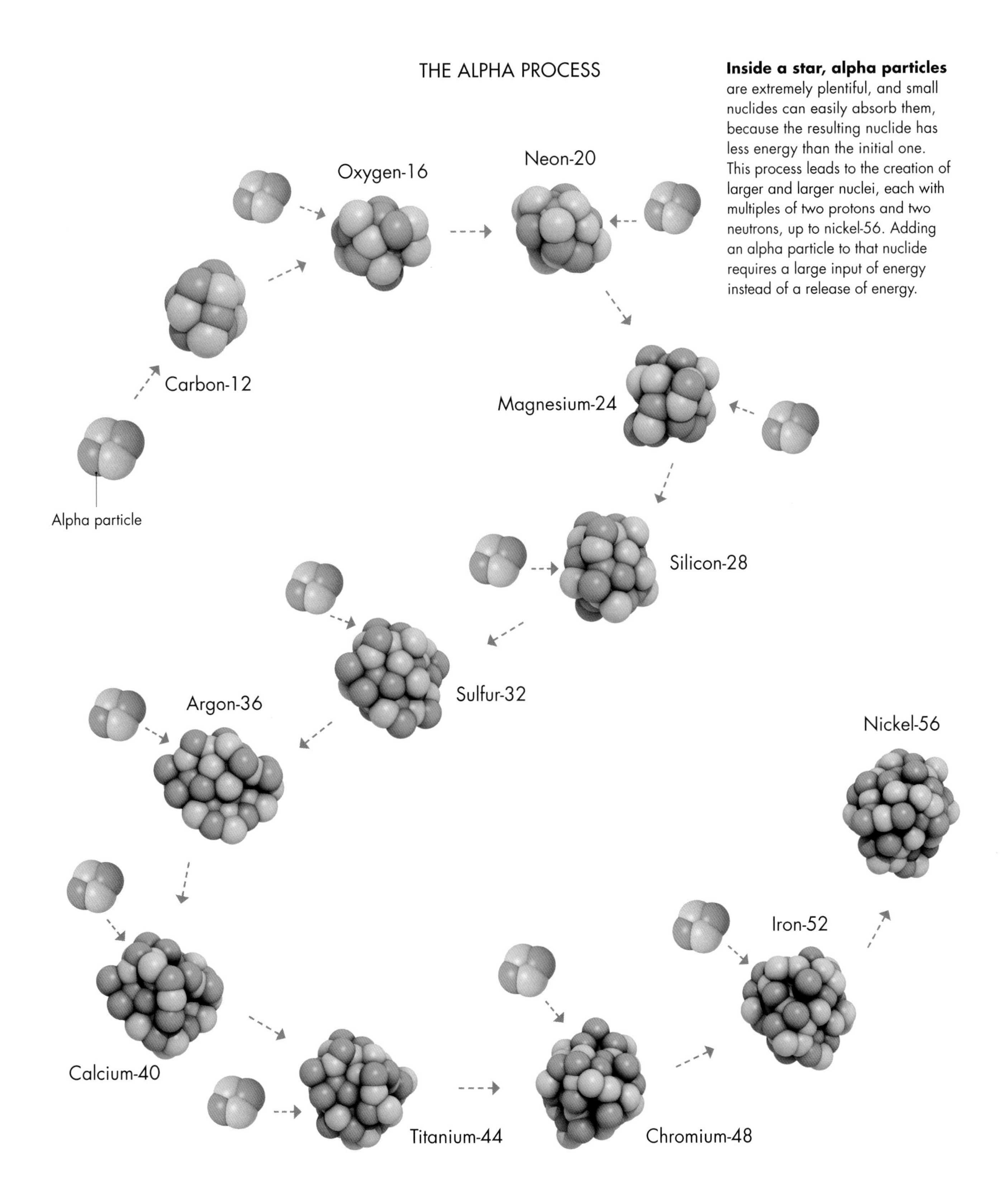

Inside a star, alpha particles are extremely plentiful, and small nuclides can easily absorb them, because the resulting nuclide has less energy than the initial one. This process leads to the creation of larger and larger nuclei, each with multiples of two protons and two neutrons, up to nickel-56. Adding an alpha particle to that nuclide requires a large input of energy instead of a release of energy.

GOING SUPERNOVA

Alongside the s-process, the alpha process continues up to the creation of the nuclide nickel-56. This nuclide requires more energy to form than it releases, unlike the lighter elements that preceded it. At this stage, then, the star has run out of fuel, and it collapses and then blows apart in a dramatic explosion called a supernova. During the supernova, more elements are created by a much more rapid version of the s-process—so rapid it is called the r-process. There are many more free neutrons available in a supernova explosion, and several can be captured by a single nucleus at once. This leads to neutron-heavy nuclei that will decay either immediately or later in successive rounds of beta decay to form any of a large number of brand new nuclides.

Two elements, beryllium and boron, are not produced by any of the processes described above, and yet are not particularly rare. Beryllium-8 (4p,4n) seems like a probable candidate for being made by the alpha process, because its nuclei are the equivalent of two alpha particles stuck together. However, it is very unstable, with a half-life of a fraction of a second. Beryllium-9, on the other hand, is stable. Along with the stable nuclides boron-11 and boron-10, it is produced by the magnificently-named process of "cosmic ray spallation." Cosmic rays are very fast-moving particles—mostly protons, alpha particles, and electrons. When these particles hit heavier nuclei during or after a supernova explosion, or even during the lifetime of a star, they can cause the nuclei to become unstable, and split, or fission (see page 160) to form these smaller, lighter nuclei. Some lithium-7 is also made this way.

Supernova explosions have another key role in the story of the origin of the elements. They scatter the whole array of nuclides, old and new, out into deep space. From there, a new generation of stars can form. The new stars may be accompanied by a spinning disk of dust and gas made of the elements created in the dead star's lifetime and at its fiery end. In many cases, the disk will coalesce into clumps that will become planets. This planetary disk is cool enough for atoms to form once again; would-be nuclei, old and new, become atomic nuclei, as electrons settle down into orbitals. The electrons fill the orbitals from the bottom up—and the best way to understand and represent this pattern of orbital filling is to arrange the elements in a table: the periodic table. Doing so also explains why certain groups of elements have similar chemical and physical properties.

THE R-PROCESS

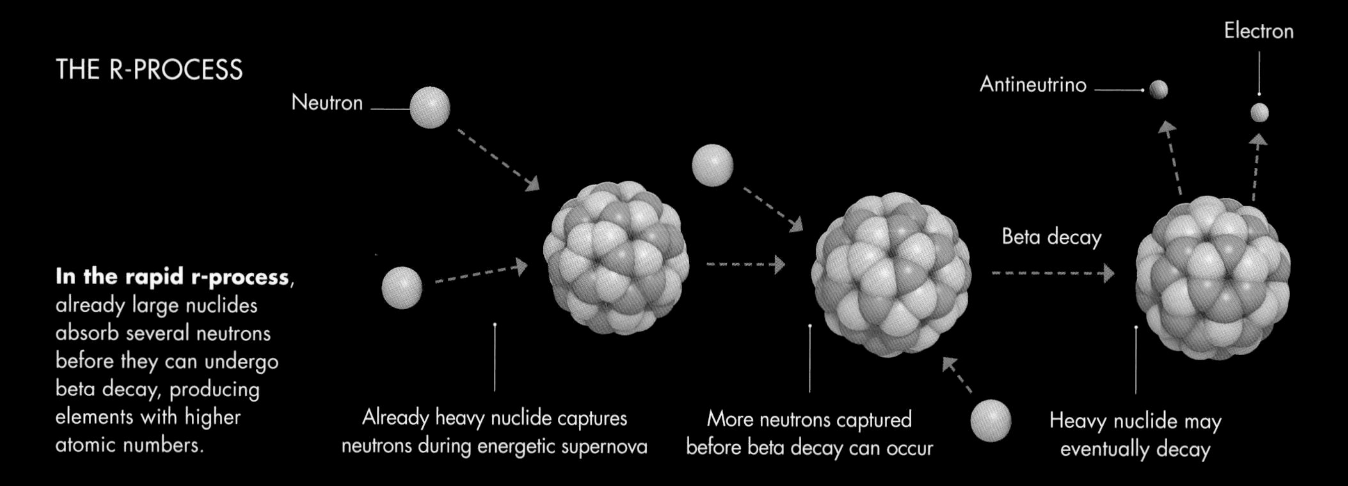

In the rapid r-process, already large nuclides absorb several neutrons before they can undergo beta decay, producing elements with higher atomic numbers.

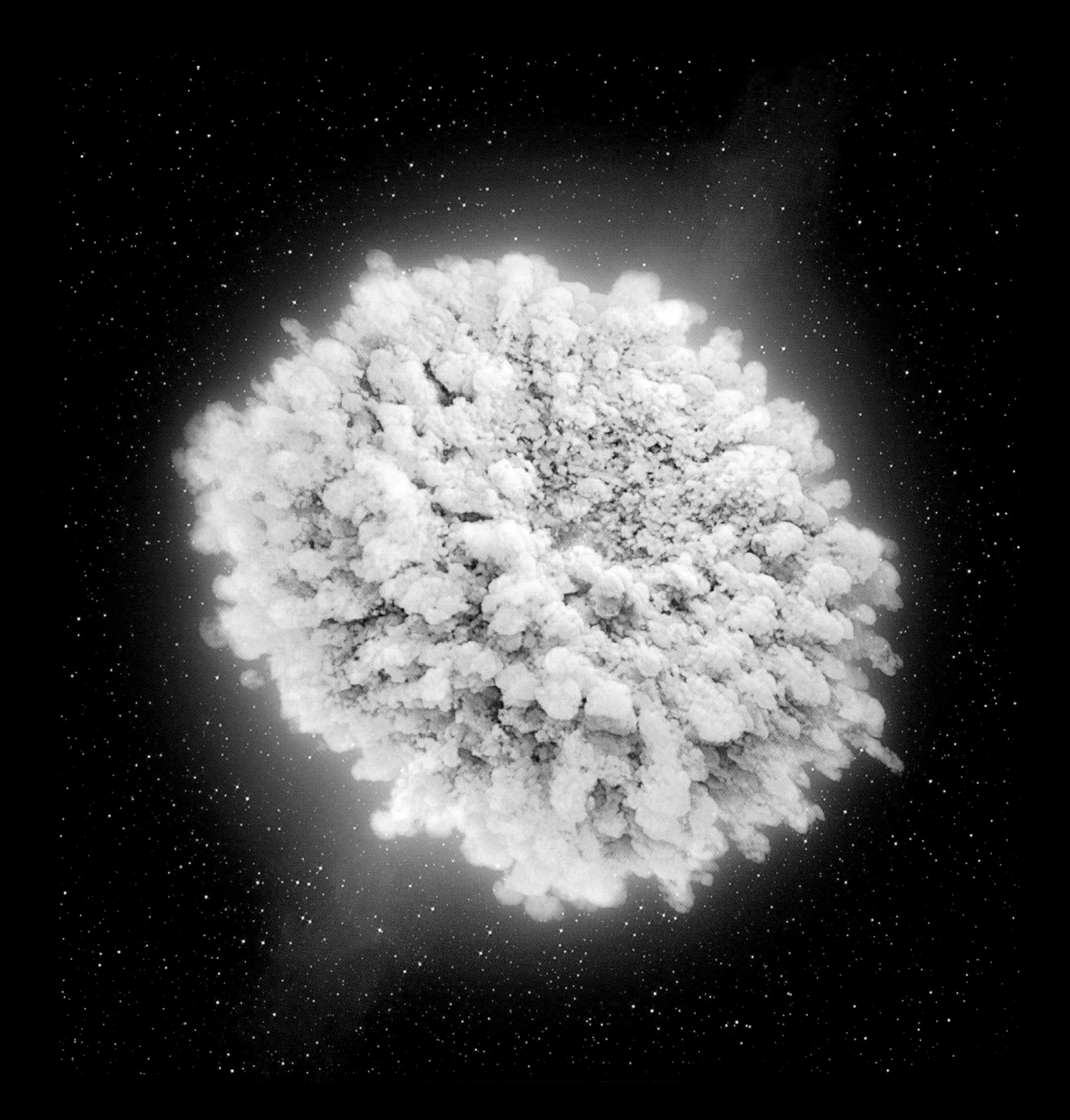

An artist's impression of a "kilonova"—an explosion produced by the collision of two neutron stars. Astrophysicists have worked out that many of the heaviest elements, including gold and platinum, are created in such collisions. A neutron star is what is left behind after a massive star has gone supernova and blown its newly-created elements out into space.

THE PERIODIC TABLE

It adorns every chemistry classroom, and is as iconic and instantly recognizable as the map of the United States or the logos that represent global brands. And yet, only a small proportion of people understand the periodic table or realize its explanatory power and its inherent beauty. The periodic table neatly expresses the links between the quantum physics of electrons in their orbitals and the chemical properties of elements in everyday life.

ADDING ELECTRONS

Each row, or period, of the periodic table represents the filling of a particular outer electron shell in the atoms of the elements that inhabit it. A shell is the collection of all the orbitals at a particular energy level (the s-, p-, d-, and f-orbitals, see pages 82–85). So the first period, at the top of the table, is home to the elements with electrons in only the first shell—in other words, those elements that have electrons at only the first energy level. That first energy level, with the lowest amount of energy, has just one available orbital: the 1s-orbital. And because each orbital can hold up to two electrons, the first period contains just two elements: hydrogen and helium. Hydrogen has one electron, while helium has two. How does this relate to the chemical properties of these elements?

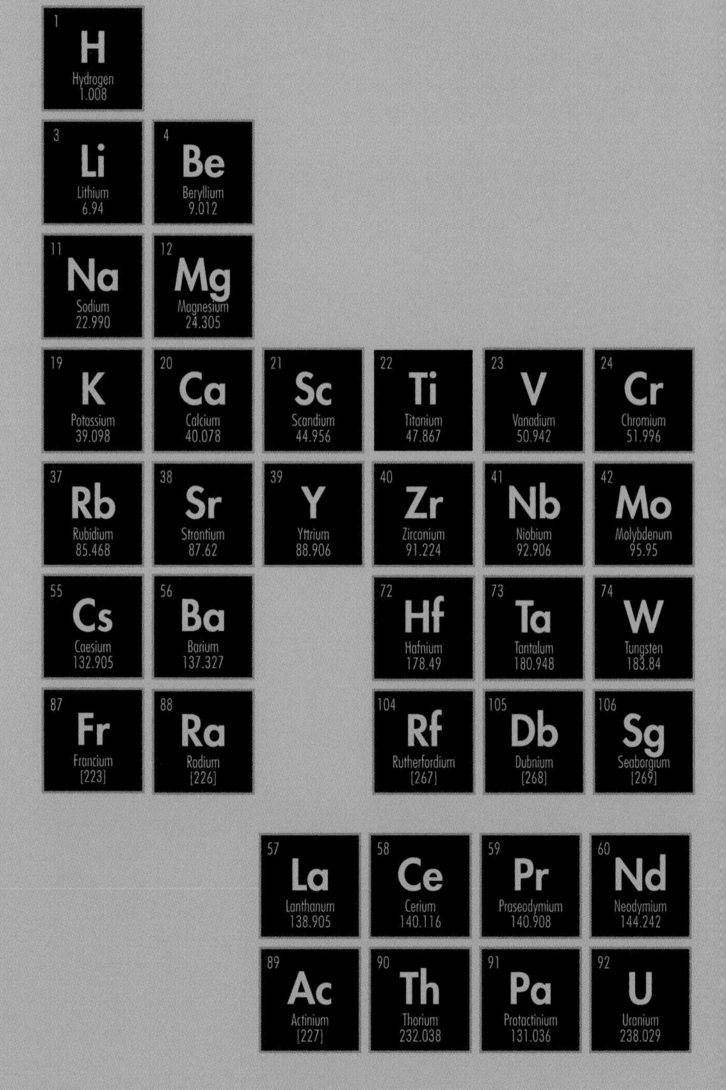

The standard form of the periodic table. Rows are called periods, and columns are called groups. Elements from any particular group have similar properties, because they have the same number of electrons in their outermost shell.

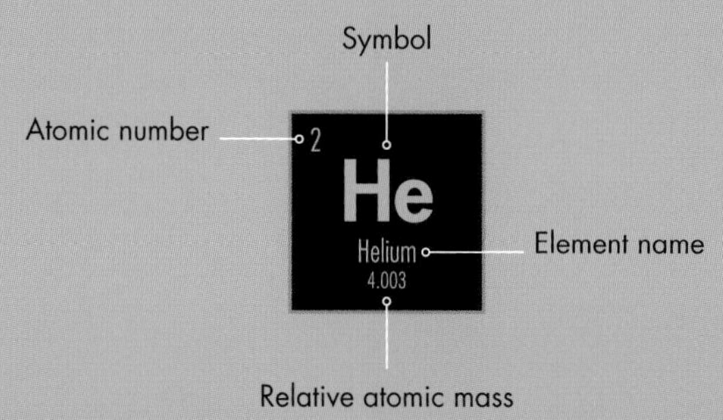

Each cell in the periodic table typically gives the name of the element in question, its symbol, its atomic number (the number of protons in the nucleus of each of its atoms), and its relative atomic mass. Also called atomic weight, relative atomic mass is the mass of a single atom of the element, in atomic mass units (daltons, see page 40). Different isotopes of the element have different numbers of neutrons, and so different masses, so this figure is an average for all the atoms of that element.

											2 He Helium 4.003
						5 B Boron 10.81	6 C Carbon 12.011	7 N Nitrogen 14.007	8 O Oxygen 15.999	9 F Fluorine 18.998	10 Ne Neon 20.180
						13 Al Aluminum 26.982	14 Si Silicon 28.085	15 P Phosphorus 30.974	16 S Sulfur 32.06	17 Cl Chlorine 35.45	18 Ar Argon 39.948
25 Mn Manganese 54.938	26 Fe Iron 55.845	27 Co Cobalt 58.933	28 Ni Nickel 58.693	29 Cu Copper 63.546	30 Zn Zinc 65.38	31 Ga Gallium 69.723	32 Ge Germanium 72.630	33 As Arsenic 74.922	34 Se Selenium 78.971	35 Br Bromine 79.904	36 Kr Krypton 83.798
43 Tc Technetium [98]	44 Ru Ruthenium 101.07	45 Rh Rhodium 102.906	46 Pd Palladium 106.42	47 Ag Silver 107.868	48 Cd Cadmium 112.414	49 In Indium 114.818	50 Sn Tin 118.710	51 Sb Antimony 121.760	52 Te Tellurium 127.60	53 I Iodine 126.904	54 Xe Xenon 131.293
75 Re Rhenium 186.207	76 Os Osmium 190.23	77 Ir Iridium 192.217	78 Pt Platinum 195.084	79 Au Gold 196.967	80 Hg Mercury 200.592	81 Tl Thallium 204.38	82 Pb Lead 207.2	83 Bi Bismuth 208.980	84 Po Polonium [209]	85 At Astatine [210]	86 Rn Radon [222]
107 Bh Bohrium [270]	108 Hs Hassium [269]	109 Mt Meitnerium [278]	110 Ds Darmstadtium [281]	111 Rg Roentgenium [280]	112 Cn Copernicium [285]	113 Nh Nihonium [286]	114 Fl Flerovium [289]	115 Mc Moscovium [289]	116 Lv Livermorium [293]	117 Ts Tennessine [294]	118 Og Oganesson [294]

61 Pm Promethium [145]	62 Sm Samarium 150.36	63 Eu Europium 151.964	64 Gd Gadolinium 157.25	65 Tb Terbium 158.925	66 Dy Dysprosium 162.500	67 Ho Holmium 164.930	68 Er Erbium 167.259	69 Tm Thulium 168.934	70 Yb Ytterbium 173.045	71 Lu Lutetium 174.967
93 Np Neptunium [237]	94 Pu Plutonium [244]	95 Am Americium [243]	96 Cm Curium [247]	97 Bk Berkelium [247]	98 Cf Californium [251]	99 Es Einsteinium [252]	100 Fm Fermium [257]	101 Md Mendelevium [258]	102 No Nobelium [259]	103 Lr Lawrencium [262]

Potassium is in Group 1 of the periodic table, so it has a single outer electron. This makes it extremely reactive—so much so that it reacts explosively with a single drop of water.

Chemical reactions have nothing to do with nuclei and everything to do with electrons. They involve the swapping and sharing of electrons between atoms—something that is explored in more detail in chapter four. Most chemical reactions result in atoms attaining a state in which their outer shell is filled with electrons. An atom with a full outer shell is in a stable, low-energy state. Elements whose atoms naturally have filled shells are therefore unreactive, while the atoms of other elements achieve a full shell by losing or gaining electrons in chemical reactions. Elements that attain a full outer shell by easily losing electrons readily form positive ions, while those that attain a full shell by gaining electrons readily form negative ions.

The atoms of elements down the right-hand side of the table, in the last column, all have full outer shells. As a result, those elements are chemically stable, because it would take a huge amount of energy to add or take away electrons from a full shell. Those elements take part in very few chemical reactions, and only in extreme situations. The columns of the table are called groups, and those unreactive elements at the right-hand side make up Group 18. This group is also called the "noble gases"—meaning noble as in incorruptible.

At the extreme left-hand side of the table—in Group 1, the alkali metals—are elements whose atoms have just one electron in their outer shell. An atom of a Group 1 element will easily lose its lone outer electron to attain a stable, full-shell state, but it is then no longer an atom: it is a positive ion (having lost one negatively-charged electron). The fact that the electron is so easily lost means that these elements are extremely reactive. Add pure potassium (K) to water, for example, and the potassium atoms will attain a full outer shell by donating their outer electrons to water molecules in a fiery display. Similarly, atoms of the elements in Group 2 are extremely reactive—but not so much as those in Group 1, because they have to lose two electrons before they realize that full outer shell.

Together, Groups 1 and 2 make up what is called the s-block of the periodic table (see page 82). In the outer shell of their atoms, electrons are only present in an s-orbital. Helium could be in the s-block, because its two electrons are both in an s-orbital. But for the purposes of the periodic table, its status as an atom with a filled shell is more important, so it is placed in with the noble gases in Group 18. Likewise, hydrogen is a little out of place in the periodic table, placed as it is with the alkali metals in Group 1. With just one electron, it is as easy for a hydrogen atom to receive an extra electron to make a full shell of two electrons, as it is for it to lose one. As a result, hydrogen could just as easily be in the penultimate group, next to helium. So the first period of the periodic table is not completely representative of the trends in the rest of the table.

BEYOND THE FIRST ELECTRON SHELL

The fact that the periods become longer after the first period, and the periodic table is wider, is testament to the fact that there are more orbitals available at higher-energy levels. In the second shell (in Period 2), for example, there is the 2s orbital plus three 2p-orbitals—room for a total of eight electrons. That is why there are eight elements in Period 2. At the right-hand end of Period 2 is the element neon (Ne). Neon has electrons in all those orbitals, plus the two electrons in the 1s-orbital below—a total of ten electrons. It is no surprise, then, that neon's atomic number is 10.

In the next row, Period 3, there are also eight slots available in the outer shell: two in the 3s-orbital and two in each of the three 3p-orbitals. The element at the right-hand end of Period 3 is argon (Ar), and its atomic number is eight more than neon's: 18 (2 + 8 + 8). The section of the periodic table containing elements whose outermost electrons are in p-orbitals is called the p-block (see page 83).

In Period 4, a new type of orbital becomes available: the d-orbital. (The fourth shell's d-orbital actually belongs to the third shell, so it is a 3d-orbital). There are five d-orbitals in any shell that has them. So the fourth shell has a total of eighteen slots to fill: a total of eight in the 4s- and 4p-orbitals and another ten in the five 3d-orbitals. That is why the width of the periods suddenly switches from eight to eighteen. The element at the right-hand end of Period 4 is krypton (Kr), with an atomic number eighteen more than that of argon: 36 ($1s^2\,2s^2\,2p^6\,3s^2\,3p^6\,3d^{10}4s^2\,4p^6$). Period 5 has the same number of elements—a total of eighteen—with the noble gas at the extreme right end being xenon (Xe), with an atomic number of 54, and 54 electrons: ($1s^2\,2s^2\,2p^6\,3s^2\,3p^6\,3d^{10}4s^2\,4p^6\,4d^{10}\,5s^2\,5p^6$)—eighteen more electrons than krypton. The section of the periodic table that corresponds to elements whose outer electrons are in d-orbitals is called the d-block (see page 84).

In Period 6, yet another type of orbital becomes available: the f-orbital. There are seven f-orbitals in each period that has them, so the width of the periodic table should grow yet again, to be 32 elements wide ($s^2\,p^6\,d^{10}\,f^{14}$). There are versions of the table that have that full width. In the standard version of the table, however, the f-block is shown separately, under the main table (see page 85).

The f-block contains elements from two periods: Period 6 and Period 7. The heaviest (nearly) stable element that occurs naturally, uranium, is found in Period 7. All the remaining "transuranium elements," all of them in Period 7 (but not all in the f-block), have been made artificially, in particle accelerators, by firing neutrons at other heavy elements, or by colliding heavy nuclei together. The noble gas at the end of Period 7 (back in the main table) is oganesson (Og), the heaviest element so far discovered or created. It has an atomic number of 118 (2 + 8 + 8 + 18 + 18 + 32 + 32). The only oganesson isotope so far created, oganesson-294, has a half-life of less than one-thousandth of a second.

THE S-BLOCK

The first two groups of the periodic table, Groups 1 and 2, make up the s-block, because the outermost electrons of the atoms of elements in these groups are in s-orbitals. Group 1, the alkali metals, includes not only the familiar elements sodium (Na), lithium (Li), and potassium (K), but also the less familiar elements rubidium (Rb), cesium (Cs), and francium (Fr). In their pure form, all these metals are extremely reactive, because they easily lose the single electron in their outermost shell to form positive ions—for example, Na^+ and Cs^+. As a result, these metals are never found in their pure state in nature, existing instead bound to negative ions, such as chloride (Cl^-). Group 2, the alkaline earth metals, also contains some familiar elements: magnesium (Mg) and calcium (Ca). Others are less familiar: beryllium (Be), strontium (Sr), barium (Ba), and radium (Ra). The atoms of these elements also easily form positive ions, but in this case, they need to lose two electrons to attain a full outer shell. So the ions are doubly-charged; for example, Ba^{2+} and Ca^{2+}. Again, these elements are reactive, and are only found naturally in compounds, bound to negative ions. Hydrogen is in the s-block, but is not an alkali metal. Note that helium is included in the s-block, despite the fact that it does not belong to either group 1 or 2. This is because its only two electrons are in an s-orbital.

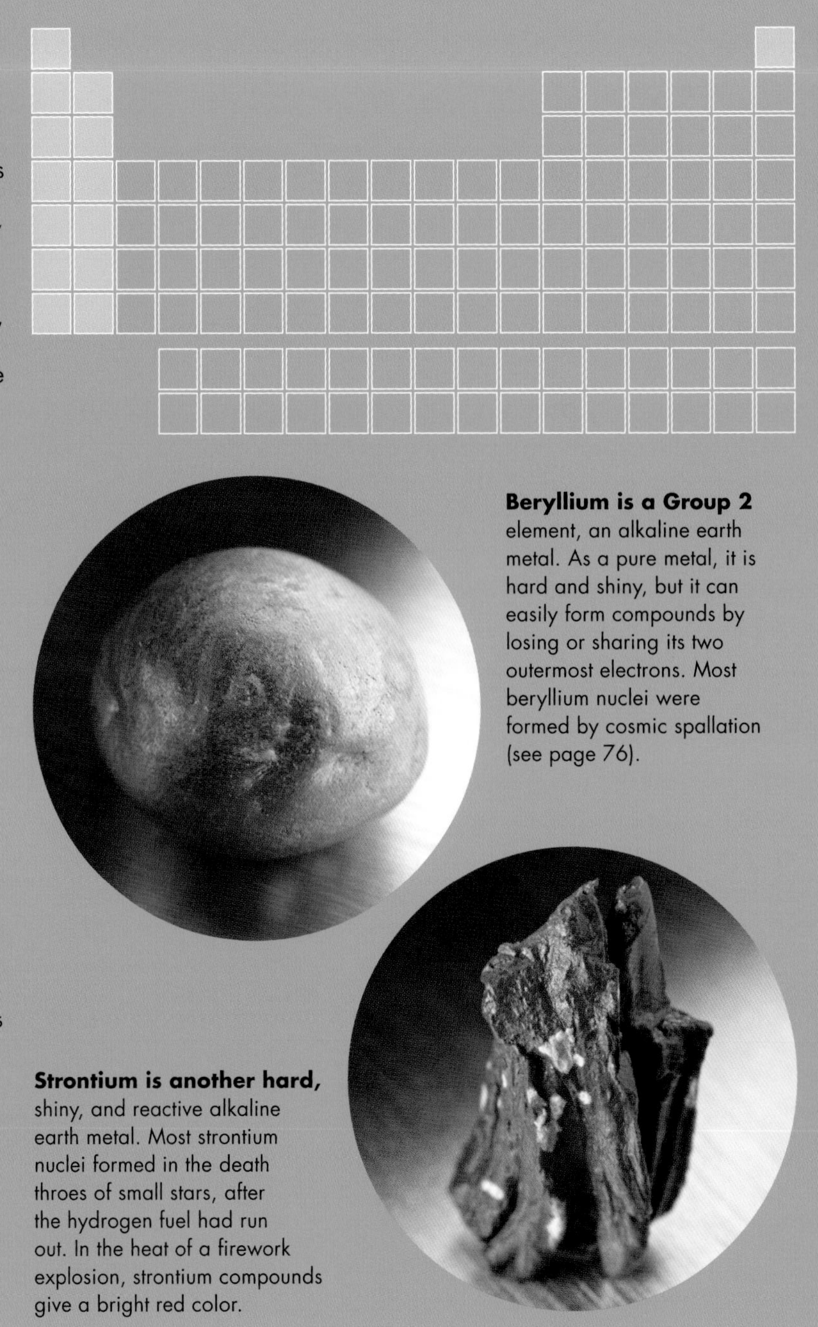

Beryllium is a Group 2 element, an alkaline earth metal. As a pure metal, it is hard and shiny, but it can easily form compounds by losing or sharing its two outermost electrons. Most beryllium nuclei were formed by cosmic spallation (see page 76).

Strontium is another hard, shiny, and reactive alkaline earth metal. Most strontium nuclei formed in the death throes of small stars, after the hydrogen fuel had run out. In the heat of a firework explosion, strontium compounds give a bright red color.

THE P-BLOCK

The elements of the p-block, which make up groups 13 to 18, are a varied bunch. They include metals, such as aluminum (Al), tin (Sn), and lead (Pb); semimetals or metalloids, such as silicon (Si) and germanium (Ge); nonmetals such as carbon (C), oxygen (O), and sulfur (S); and the noble gases in Group 18. Helium is not really a p-block element, because it has no p-orbitals, but it sits at the top of Group 18 because it has a full shell. The reason for the variety in chemical behaviors in the p-block is again to do with the likelihood of losing or gaining electrons to attain a full outer shell. Aluminum in Group 13, for example, will lose its three outermost electrons (s^2 p^1) to become a triply-charged ion (Al^{3+})—classic metal behavior. Chlorine in Group 17 will easily gain one electron to become a negative ion (Cl^-)—classic nonmetal behavior.

Between these two extremes, elements such as oxygen (Group 16) can form the doubly-charged ion O^{2-} to attain a full outer shell, but it will also form bonds that involve sharing electrons with other atoms, achieving the same effect (discussed in chapter four). At the heart of the p-block are semimetals, such as boron and silicon. These rarely form ions, and instead enter almost exclusively into electron-sharing bonds. Carbon is a special case, not least because it is the basis of life on Earth. This special case will also be explored in chapter four.

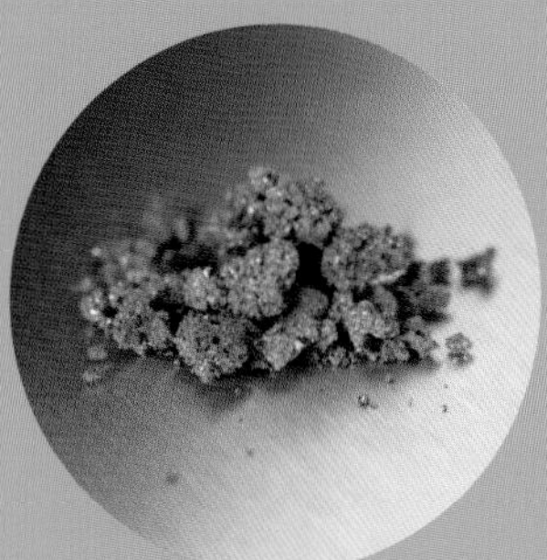

Boron, in Group 13, is classed as a metalloid. It is shiny like a metal, but it is brittle and does not conduct electricity well, while true metals can be beaten and drawn into shape and are good electrical conductors.

Silicon is a metalloid, like boron. It is also the quintessential semiconductor, with electrical conductivity between that of a conductor and an insulator. This makes it a key player in the modern electronics industry (see page 140).

Aluminum is only one place below boron, but it is a true metal. Its outermost electrons are farther from the nucleus and so more readily "delocalized" from the atom. This allows for aluminium atoms to engage in metallic bonding (see page 107).

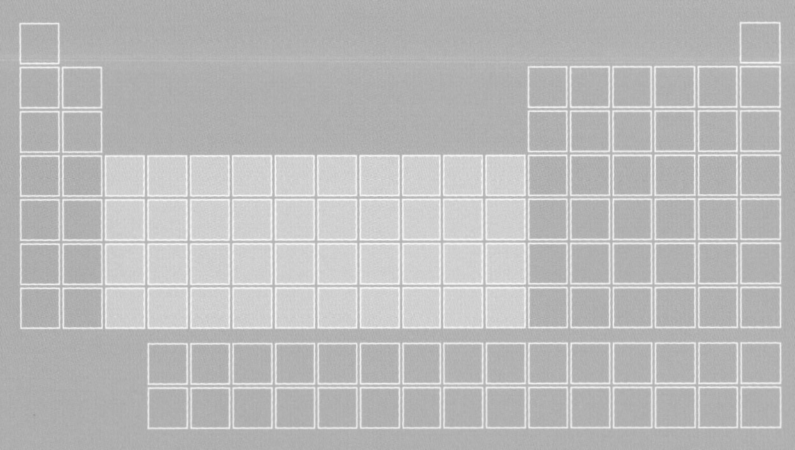

THE D-BLOCK

Elements in the d-block, with their outermost electrons in d-orbitals, are all metals; collectively, they are known as transition metals. The familiar metals of everyday life are here—including iron (Fe), copper (Cu), gold (Au), and silver (Ag)—as well as less common ones, such as niobium (Nb), osmium (Os), and ruthenium (Ru). While the metals of the s-block easily lose their electrons to become positive ions, it is not so simple for the d-block metals. Transition metal atoms can have anything between three (s^2 d^1) and twelve (s^2 d^{10}) electrons in their outer shell, and there are many different ways in which they can attain full outer shells (or the equivalent by sharing electrons). A more general definition of a metal applies. A metal is an element that is hard, but also malleable (can be beaten into shape), and ductile (can be drawn out into a wire). These metallic properties are the result of the bonds between the atoms of a metal, which are discussed further in chapter four.

Ruthenium's outermost electrons are $4d^7$ and $5s^1$. The electrons in the 4d-orbitals have more energy than those in the 5s-orbitals. This mixing of energy levels, characteristic of transition metals, is due to the fact that a d-orbital at one energy level has more energy than the s-orbital of the energy level above (here, 5), which therefore fills first.

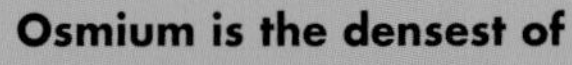

Osmium is the densest of all the elements. A block the size of a domestic washing machine would have a mass of about 20 tons. Most osmium nuclei are formed when neutron stars collide (see page 77).

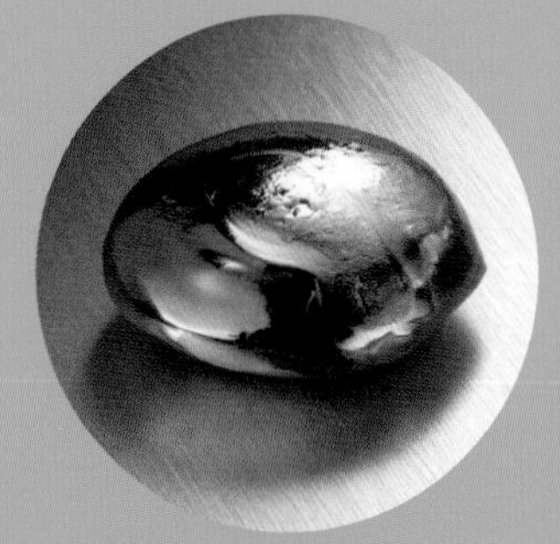

Niobium is soft, but combined with other metals, it produces extremely hard and heat-resistant alloys with applications in aerospace. The rocket nozzle of the Lunar Service Module used in the Apollo missions to the Moon was made of a niobium alloy.

THE F-BLOCK

Normally separate below the main table, the f-block contains two rows—the section of each of Period 6 and Period 7 that corresponds to elements with their outermost electrons in f-orbitals. All of these elements are metals—although the definition and classification breaks down with some of the heaviest elements, which do not exist long enough for their chemical properties to be investigated. The first row of the f-block is also referred to as the lanthanides, because it starts with the element lanthanum (La). All the lanthanides are "rare earth metals," most of which have uses in modern electronics and are used to make strong magnets. The second row is referred to as the actinides, because it begins with the element actinium (Ac). The elements beyond uranium (U) are all too unstable to exist naturally in significant quantities, although the elements neptunium (Np), plutonium (Pu), americium (Am), and curium (Cm) are stable enough, once created in nuclear reactors or particle accelerators, to be useful in a number of niche applications. From then on, however, the elements exist too fleetingly to find any real applications beyond scientific study. The most stable isotope of the heaviest actinide, lawrencium (Lw), has a half-life of just ten hours.

Samarium is a fairly hard, shiny metal. It is a lanthanide, found in the first line of the f-block of the periodic table. Like several other lanthanides, its main use is in making strong magnets.

Holmium's outermost electrons are $4f^{11}\ 6s^2$. As for transition metals (see ruthenium, far left) this mixing of energy levels is shared by all lanthanides and is due to the fact that an f-orbital at one energy level has more energy than the s-orbital of the energy level above (here, 5), which therefore fills first.

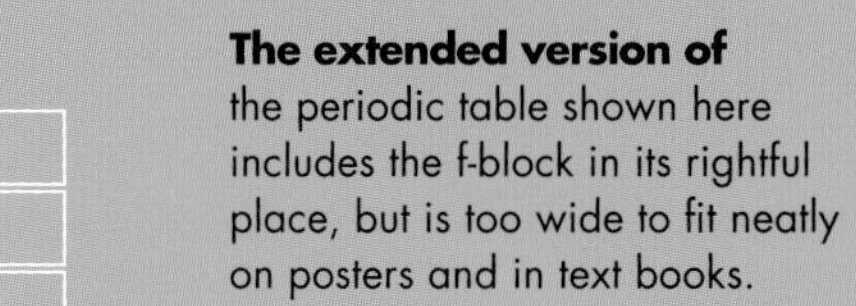

The extended version of the periodic table shown here includes the f-block in its rightful place, but is too wide to fit neatly on posters and in text books.

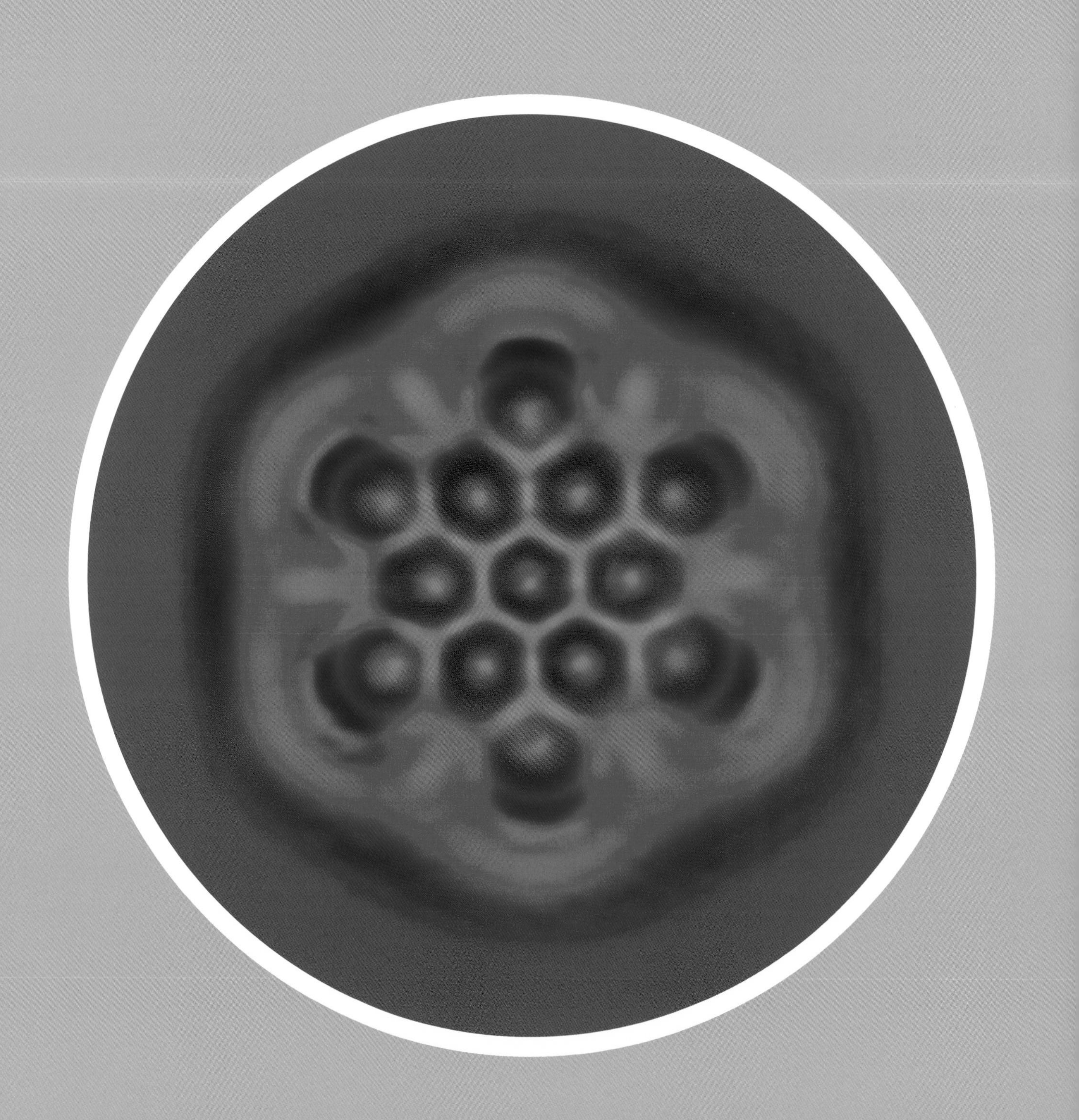

CHAPTER 4
ATOMS TOGETHER

Atomism encourages a mechanistic view of the world—one in which the familiar properties of matter at everyday scales can be explained by the behavior of unimaginably large numbers of unimaginably small particles. Those particles are in constant movement and are able to join and separate. The mechanistic view originally treated atoms as solid, impenetrable balls—and it was successful even then. But when enhanced by the modern understanding of the inner workings of the atom, it is very successful indeed.

A molecule of the compound hexabenzocoronene is shown in a false-color image produced by atomic force microscopy (see page 126). This molecule is composed of forty-two carbon atoms and eighteen hydrogen atoms, held together by covalent bonds.

MATTER AS PARTICLES

One of the most basic and powerful pieces of evidence in favor of the idea that matter is made of particles is the fact that it can explain the existence of solids, liquids, and gases, and in particular how matter can change between these states. The particles of which a solid, liquid, or gaseous substance is made can be atoms, molecules, or ions.

SOLIDS: PARTICLES STAYING PUT

A solid is rigid because its particles are joined, or bonded, together. The bonds between the particles keep them in fixed positions. Imagine those bonds as rubber bands, stronger in some solids than others—the equivalent of some having thicker or more resilient rubber bands. Although the particles are in fixed positions, they are moving; they constantly vibrate to and fro at random, in all directions. If the particles are molecules, there are also bonds between the atoms of which the molecules are made. These bonds are able to flex up and down or to and fro (by shortening and lengthening).

To get an idea of the nature of these interatomic bonds, imagine that in addition to the rubber bands holding the molecules together, there are also smaller (but typically stronger) rubber bands holding the constituent atoms together. So, for example, a water molecule in an ice cube will be vibrating randomly to and fro as a whole, but its constituent atoms (two hydrogens and an oxygen) will also be vibrating up and down within each molecule, and moving toward and away from each other. A lot of movement is going on at the atomic scale.

Model of the crystal structure of selenite. Calcium ions (calcium atoms with an electric charge) are the large gray spheres. Sulfur is yellow; oxygen is red; and the hydrogen atoms in the water molecules are the smaller pale gray spheres.

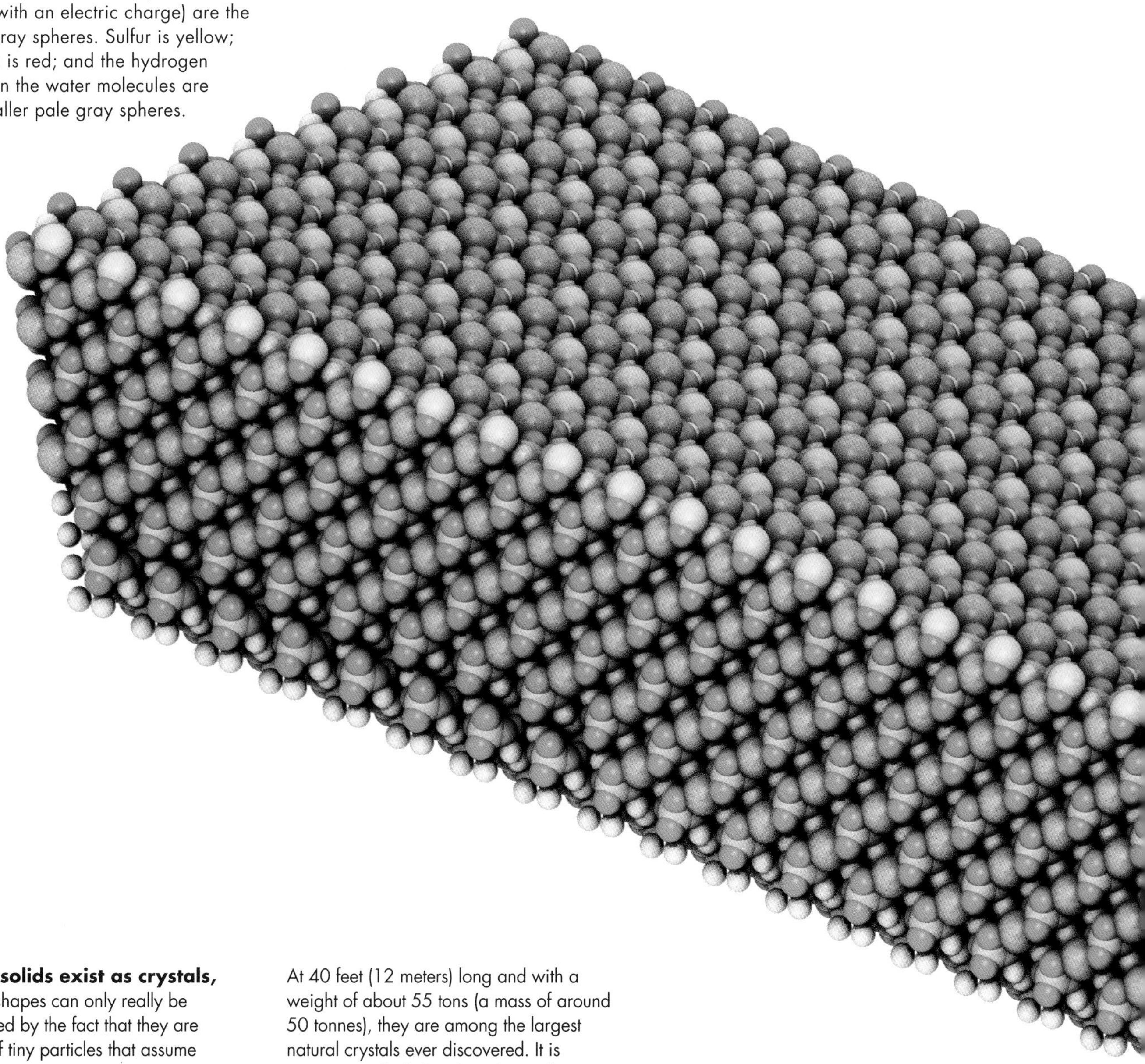

Many solids exist as crystals, whose shapes can only really be explained by the fact that they are made of tiny particles that assume regular, repeating arrangements. Incredible crystals of the mineral selenite ($CaSO_4 \cdot 2H_2O$) can be found in the Cave of the Crystals in the Naica Mine, in Chihuahua, Mexico (seen left).

At 40 feet (12 meters) long and with a weight of about 55 tons (a mass of around 50 tonnes), they are among the largest natural crystals ever discovered. It is amazing to think that these huge objects are made of only four types of atom—and that the ordered arrangement of those very small particles gives rise to the elegant, regular shape of the whole crystal.

Any moving object has energy by virtue of its motion. The amount of this "kinetic energy" depends upon the mass of the object and how fast it is moving. Being so minuscule, then, each particle of a solid has only tiny amounts of kinetic energy. But because of the large numbers of particles present, the total amount of "internal energy" the solid possesses is significant. This internal energy is related to the solid's temperature. In fact, the temperature of any substance—solid, liquid, or gas—is determined by the average kinetic energy per particle. Increase a substance's temperature, by giving it more internal energy, and you increase the average speed of the particles of which it is made. So the higher the temperature, the more vigorously the particles in a solid vibrate.

There are many ways to increase the internal energy of a solid—to raise its temperature. You could, for example, put the solid in a hot oven. In that case, the particles of hot air surrounding the solid fly around at high speed—they possess a lot of kinetic energy. When they hit the surfaces of the solid, they transfer some of their kinetic energy to the particles at the surface, and this will (literally) have a knock-on effect. Those particles at the surface, now vibrating more vigorously, pass on some of that extra kinetic energy to their neighbors just beneath the surface. Gradually, the extra energy is shared throughout the whole solid. This is the atomic-scale story of what we know as conduction, which is the transfer of heat within a substance.

THERMAL DECOMPOSITION

At high temperatures, many solids that are made of molecules will decompose instead of melt. In other words, the bonds between the atoms that make up the compound break when they are heated.

A good example is the compound mercury (II) oxide, shown here. The compound exists as an orange solid that consists of mercury ions (Hg^{2+}) and oxygen ions (O^{2-}) held together in a regular crystal structure. Heating the mercury (II) oxide to about 950°F (500°C) boosts the kinetic energy of the ions and causes them to break apart. The result is pure mercury vapor, the shiny metal you can see condensing on the inside of the test tube, and pure oxygen, which is an invisible gas. Many other compounds will decompose like this when heated, and can never exist in liquid form. All pure elements, however, can exist as solids, liquids and gases.

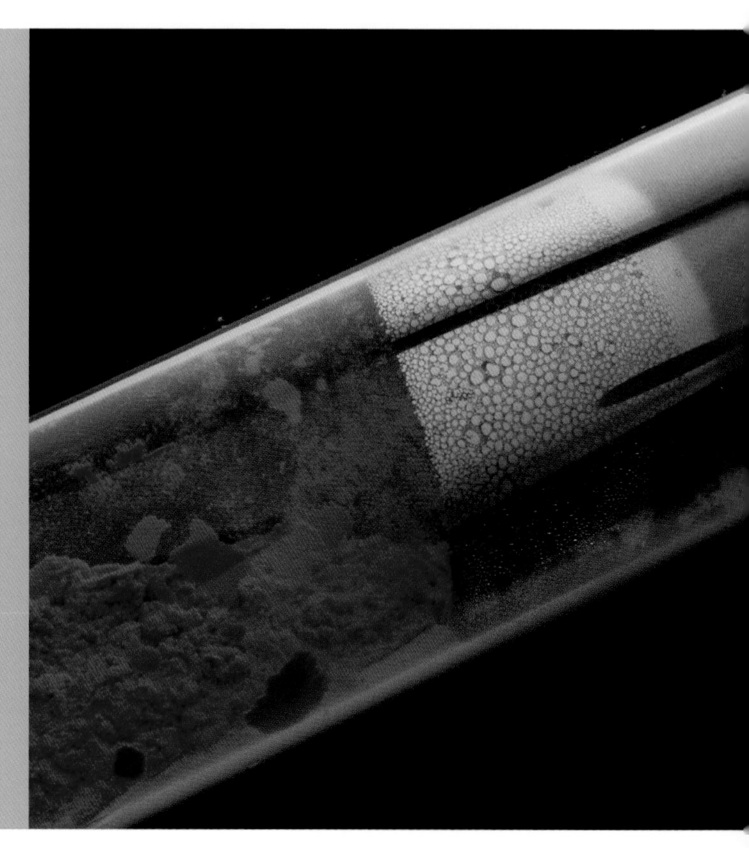

At just 86°F (30°C), atoms of the element gallium have enough energy to break free from the bonds that hold them in a solid crystal and become liquid. This means gallium will slowly melt if placed on a warm hand.

Take the solid out of the oven into the cooler air, and the reverse occurs. Particles at the surface of the solid, vibrating vigorously with more kinetic energy (on average) than the particles of the air around them, will transfer some of their energy to those air molecules. The more energetic particles within the body of the solid now transfer some of their energy to the ones at the surface, and these in turn continue to lose energy to the air molecules. And so the solid's temperature gradually decreases until it matches that of the ambient air.

In many cases (but not all, see box, left), when you increase the temperature of a solid sufficiently to its melting point it will become liquid. Its particles then have so much kinetic energy that they can break free from each other; the rubber bands are stretched to breaking point.

LIQUIDS: PARTICLES ON THE MOVE

When a solid melts—that is, when the strong bonds between the particles break—there is still an attractive force pulling the particles together (otherwise they would simply fly apart, completely separate from each other). But the particles are no longer in fixed positions; they are now able to move over and past one another. This is why a liquid can flow, and why it assumes the shape of any container that holds it.

As for solids, the temperature of a liquid is related to the average kinetic energy of its particles. Some particles of a liquid move more slowly than the average; others travel faster. At the liquid's surface, some of the faster-moving particles have enough kinetic energy to break free. Those energetic particles leave the liquid and become part of the air. This is the atomic-scale story of what we call evaporation. But only those particles with higher than average energy have enough speed to leave the surface, so the average kinetic energy of the particles left behind decreases. The liquid's temperature goes down. This process is known as "evaporative cooling." It is why sweating, for example, is an effective way for the human body to cool itself. The water from sweat evaporates and cools the skin. The rate at which a liquid evaporates depends upon the liquid's temperature and also the pressure of the air around the liquid—and, in the case of sweat, on how much water is already in the air. Reduce the air pressure, increase the body's temperature, or reduce the humidity, and the rate of evaporation increases.

Above a certain temperature (assuming the particles do not undergo thermal decomposition, see page 90), or below a certain pressure, all of the particles of a liquid will have enough energy to break free from each other. At that point the liquid boils, to become a gas. Because no element can decompose, every element can exist as a gas. At normal atmospheric pressure, molten iron boils at 5182°F (2862°C), while liquid nitrogen boils at a chilly -423.182°F (-252.879°C).

GASES: PARTICLES FLYING FREE

Whether the particles of a gas are individual atoms, molecules, or ions, they fly around freely at high speeds. They collide with each other and with any surfaces they meet. When particles collide with a surface, they exert pressure on it—one example is when a balloon stays inflated. The effervescence of a pressurized bottle of a carbonated drink, the air pressure that inflates a tire, a firework shooting into the sky—these, too, are caused by countless collisions of tiny, fast-moving particles. The more energy the particles have, the faster they move

MASS AND ENERGY

Helium atom, mass 4 daltons

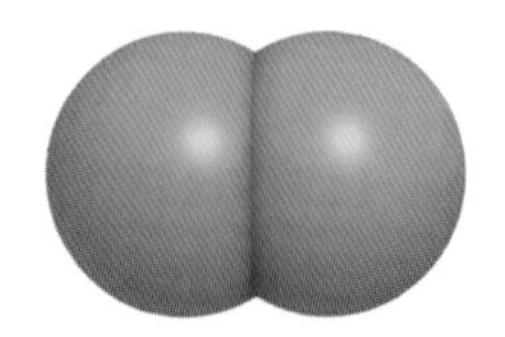

Oxygen molecule, mass 32 daltons

Hydrogen molecule, mass 2 daltons

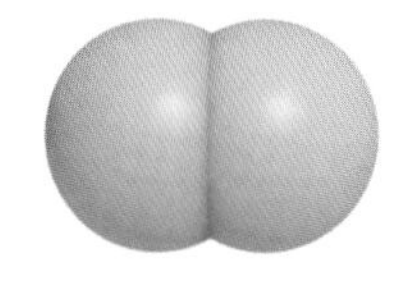

Nitrogen molecule, mass 28 daltons

Four of the elements are found as gases in the atmosphere. Although they are the two most abundant elements in the universe, hydrogen and helium are extremely rare in the atmosphere. This is because, for the same kinetic energy, these much lighter elements move so fast that they escape Earth's atmosphere.

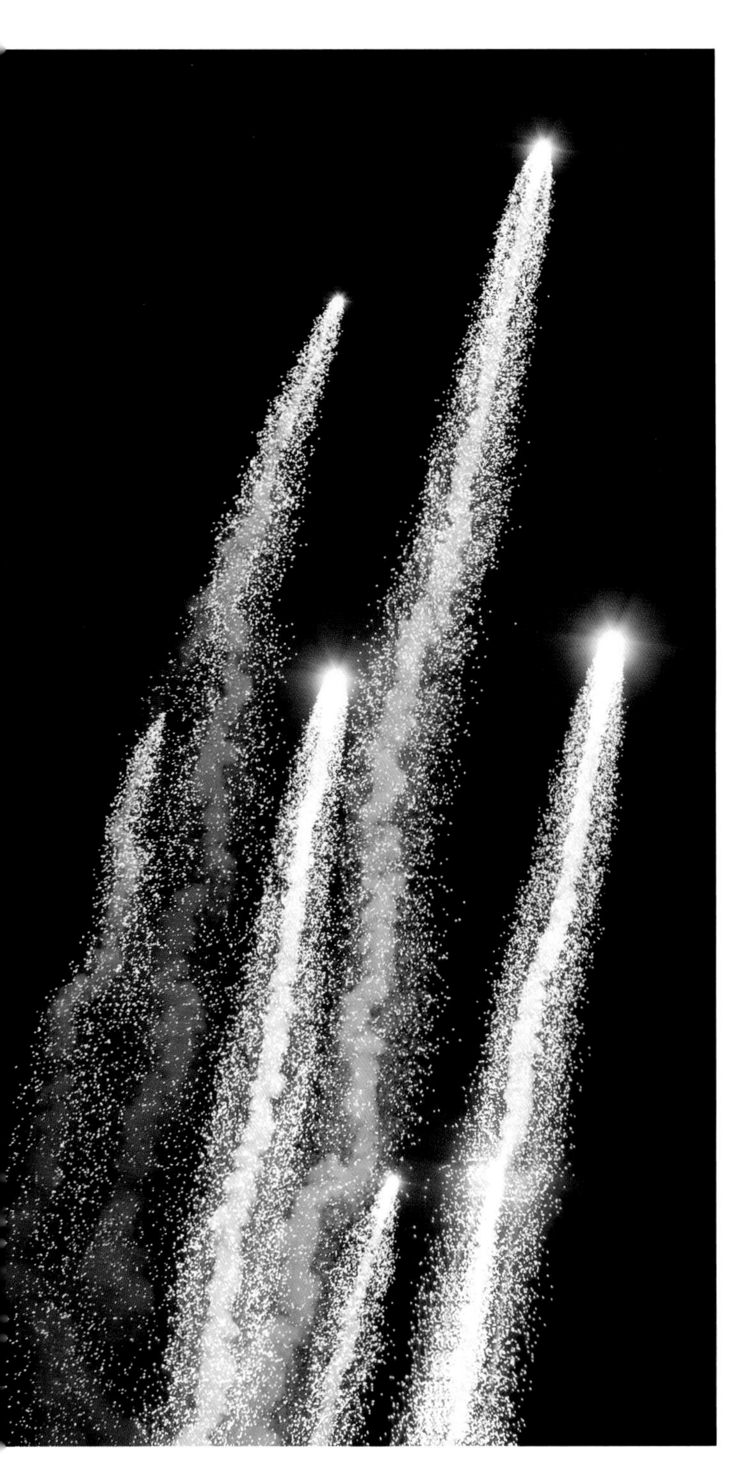

(on average). That is why increasing the temperature of a gas increases the pressure. And that, in turn, explains why the hot gases being produced in a firework expand so rapidly and accelerate the firework upward.

The gas we are most familiar with, air, is mostly made of nitrogen and oxygen. Both of these elements normally exist as diatomic molecules—N_2 and O_2—each made of two identical atoms. At room temperature, oxygen and nitrogen molecules fly around at average speeds of about 900 miles per hour (1,400 kilometers per hour). In air at room temperature, a hydrogen molecule (H_2) and a helium atom (He) have the same average energy as molecules of oxygen and nitrogen, because the particles of a gas exchange energy whenever they collide with each other. But the helium atoms and hydrogen molecules have far less mass than the nitrogen and oxygen molecules, so, for the same amount of energy, their average speed will be much greater. The average hydrogen molecule or helium atom in the air is moving around five times as fast as the average nitrogen or oxygen molecule—fast enough to escape Earth's atmosphere. Indeed, the atmosphere loses 105,000 tons (95,000 tonnes) of hydrogen and 1,800 tons (1,600 tonnes) of helium each year.

One other important component of air is water vapor. In just the right conditions of temperature and pressure, the vapor can condense to become tiny droplets, forming a mist. A mist is an example of an aerosol, one of a class of substances referred to as colloids.

Firework rockets accelerate rapidly upward because of the expanding exhaust gases escaping out of the bottom. The gases expand because they are hot, which makes the molecules move at high speeds and bounce off each other and the sides of the rocket.

COLLOIDS—NOT ONE THING OR ANOTHER

Most of the substances with which we are familiar are mixtures. Some are solutions, in which the smallest particles of one substance have become thoroughly and evenly mixed in with the smallest particles of another substance. Saltwater is a solution in which sodium and chloride ions are dispersed evenly among the water molecules. Surprisingly, perhaps, steel is also a solution: a solid solution. It is mostly made of iron atoms, but they are interspersed with the individual atoms of several other elements. And any mixture of gases can be considered a solution, because the gas particles will naturally disperse as the gases mix. Not all mixtures, however, are solutions; many are colloids.

In a colloid, as with a solution, the particles of one substance are dispersed evenly among those of another. But the dispersed particles are not individual atoms or ions, nor even individual molecules. Instead, they are tiny liquid droplets, tiny solid particles, or tiny pockets of gas—each one very small, but still made of billions or trillions of atoms. Mayonnaise, for example, is a type of colloid called an emulsion. It is made of tiny droplets of oil dispersed in another liquid (vinegar). Smoke is a type of colloid called an aerosol, which is made of tiny solid particles dispersed evenly in a gas (air). Steam and mist are liquid aerosols: tiny liquid (water) droplets dispersed evenly in a gas (air). Gelling agents, such as gelatin, form a type of colloid called a gel when water is added. In that case, tiny droplets of water are dispersed evenly amid the solid structure of the gelatin. Another type of colloid, called an aerogel, is similar, but here air or other gases are trapped within the solid structure, instead of water.

In solids and liquids—whether in a colloid or not—there are bonds between the particles of which they are made. In a solid that is made of ions, the ions are held together by an electrical attraction called an ionic bond. If the particles of a substance are molecules, there is another type of bond holding the atoms of each molecule together. Both these types of interatomic bonds involve the outermost electrons of the atoms.

Aerogel, pictured here, is a type of colloid in which air (a gas) is mixed with a solid (in this case silicon dioxide). It is extremely light and very insulating. The other examples are meringue (a solid foam); mayonnaise (an emulsion); steam (an aerosol); gelatin (a gel); and smoke (a solid aerosol).

Meringue
Mayonnaise
Steam
Gelatin
Smoke

ATOMS JOINING TO ATOMS

How wonderful it is that atoms can join, or bond, together. If they could not, the universe would be filled with individual atoms, bouncing off each other but remaining strictly single. There would be no chemical compounds—and certainly no life. Atomic nuclei are not involved in the making and breaking of bonds. That honor is bestowed upon the electrons—and, in particular, the electrons in the outermost shell.

The outermost electron shell of an atom is also called the valence shell. In atoms of the noble gases (see page 80), such as argon and krypton, the valence shell is completely filled; there are two electrons in each of the available valence orbitals. A filled valence shell is a "desirable" and stable state for an atom, because it will require a good deal of energy to take electrons away or to add more. This stability is the reason why the noble gases do not form bonds with other atoms. Other atoms, with valence shells that are not completely filled, can attain full-shell status either by exchanging or sharing electrons—and these two options are the basis of interatomic bonding.

IONIC BONDING

The simplest way in which an atom can attain a full valence shell is to donate one electron, or more than one electron, to another atom, or to receive one or

IONIC CRYSTAL

In a solid sample of lithium fluoride, countless trillions of ions cling together by virtue of their mutual electrical attraction. Lithium fluoride is a cubic crystal, because the ions exist in a cubic "lattice." In calcium carbonate, the ions arrange in a hexagonal pattern.

CALCIUM AND CARBONATE IONS

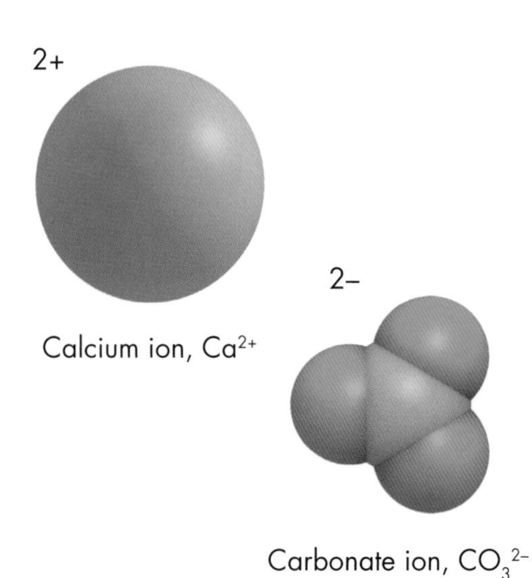

Calcium, in Group 2 of the periodic table, easily loses its two outermost electrons. When it does so, it becomes a calcium ion, with a charge of 2+. A carbon atom bonds with three oxygen atoms to form a small group that has an overall charge of 2–.

more donated electrons. So, for example, an atom of the element lithium ($1s^2\,2s^1$) can easily lose the single electron in its valence shell. If it does so, it becomes a lithium ion, Li^+ ($1s^2$). And if an atom of fluorine ($1s^2\,2s^2\,2p^5$) is nearby, it can attain a filled valence shell by accepting that electron and in the process become a fluoride ion, F^- ($1s^2\,2s^2\,2p^6$). Now two ions have been created with opposite electric charges. These strongly attract each other and cling together. The result is a compound, known as lithium fluoride (LiF), which is a solid at room temperature.

Ionic bonding between individual atoms happens exclusively between a metal and a nonmetal. Sodium (Na, a metal) and chlorine (Cl, a nonmetal) provide the best-known example, with their ions (Na^+ and Cl^-) clinging together to form the compound sodium chloride (NaCl), or common salt. But the ions involved in ionic bonding can also be "polyatomic," in other words, they may consist of more than one atom. Calcium carbonate ($CaCO_3$), the example illustrated here, is an ionic compound that consists of the ions Ca^{2+} and CO_3^{2-} (see page 96).

CRYSTAL STRUCTURE OF CALCIUM CARBONATE

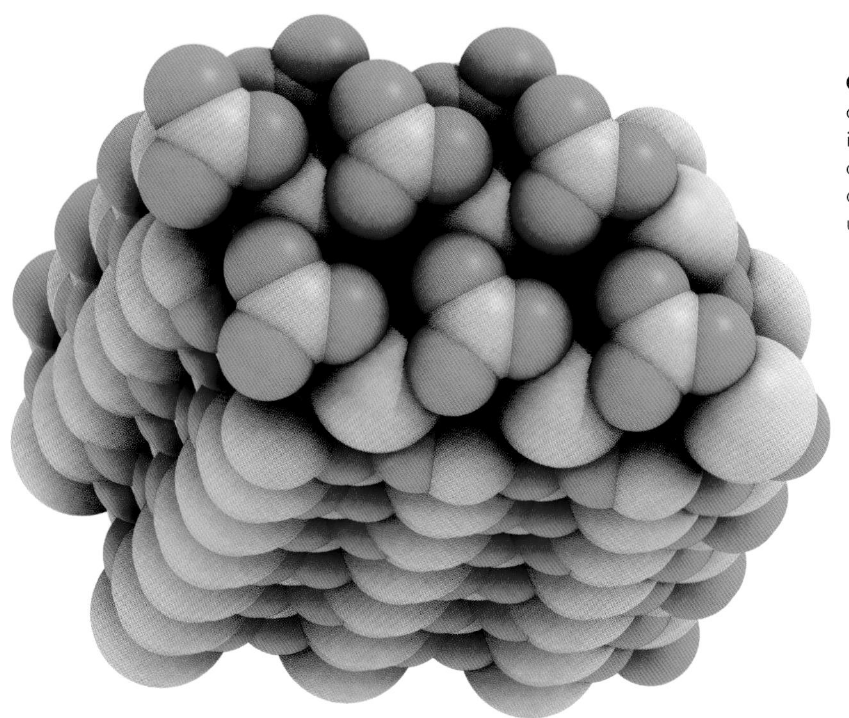

Calcium carbonate is one of the most common minerals in Earth's crust. Two kinds of calcium carbonate, with slightly different crystal structures, make up the rock called limestone.

COVALENT BONDS

The other way in which atoms can attain a full valence shell, and form a bond, is by sharing electrons. In this case, orbitals of two atoms merge, forming a new "molecular orbital" that connects and binds them. Because the two atoms share valence electrons, this type of bond is described as covalent.

The simplest covalent bond is the bond between the two atoms of a hydrogen molecule (H_2). It is formed by the overlap of the two atoms' 1s-orbitals. The merging of atomic orbitals creates a molecular orbital—and just as for an atomic orbital, it is the region where the electrons will probably be found. The orbital has a full complement of two electrons—both atoms "feel" as though they have a filled 1s orbital—so this is a desirable situation in terms of energy. The bond is short and lies along the line between the two hydrogen nuclei. Any interatomic bond that lies along the line between two nuclei is called a sigma bond. There are other types of molecular orbitals, each created by the merging or overlapping of atomic orbitals (see box on the facing page).

A molecule, then, is a self-contained object made of atoms held together by covalent bonds. Every water molecule is made of two hydrogens covalently bonded to an oxygen atom. The same is true of the two oxygen atoms bonded to a carbon atom in carbon dioxide (CO_2). The wax of a candle consists of molecules that each contain about thirty carbon atoms and about sixty hydrogen atoms, all bonded covalently. The shape of a molecule is determined by the lengths and orientations of the bonds between the atoms, but it is also influenced by any other electrons not involved in the bonding. So in water, for example, there is a (sigma) bond between each hydrogen atom and the oxygen atom, but the oxygen atom has other electrons in its valence shell. Their presence pushes the bonds away, giving the molecule a bent shape.

WATER MOLECULE

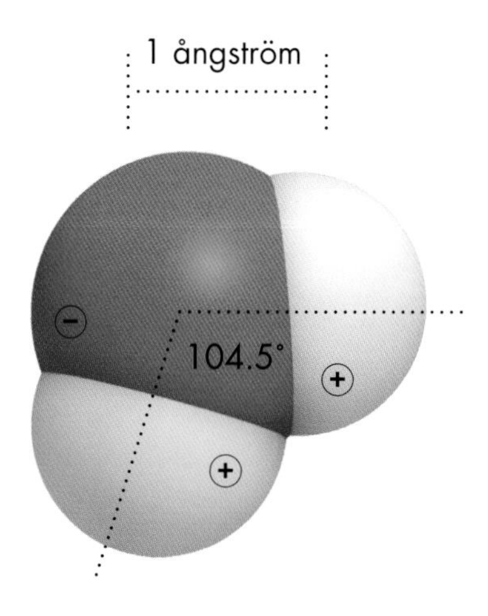

A water molecule is made of two hydrogen atoms and one oxygen atom bonded covalently. The electrons are more concentrated around the oxygen atom, which therefore has a partial negative charge, leaving the hydrogen atoms carrying a partial positive charge.

The presence of the electrons belonging to the oxygen atom in a water molecule has another consequence. It leads to a relative concentration of negative charge around the oxygen atom, while each hydrogen atom—with its lone electron committed to the bond between it and the oxygen atom—represents parts of the molecule that have a relative positive charge. This situation gives a water molecule "polarity," the reason why water is such a good solvent.

Consider what happens when salt (sodium chloride) dissolves in water (see the illustration on page 100). The sodium chloride breaks up into its constituent ions fairly easily under the influence of polar water molecules. The positively-charged sodium ions stick to the oxygen atoms of nearby water molecules, while the negatively-charged chloride ions stick to the hydrogen atoms. The same happens with a host of other ionic solids and to compounds made of polar molecules; they all dissolve well in water. Nonpolar molecules, in which there is a uniform distribution of electric charge throughout, do not dissolve well in water. Fats and oils are good examples.

MOLECULAR ORBITS

Covalent bonds that hold together molecules are formed by the overlap of atomic orbitals (see page 54). There are several possible arrangements, the most simple and strongest being a sigma bond: the head-on overlap of two orbitals. A pi-bond is formed by the side-on overlap of two p-orbitals. In an oxygen molecule (O_2), the oxygen atoms are held together by a double bond: one sigma and one pi bond.

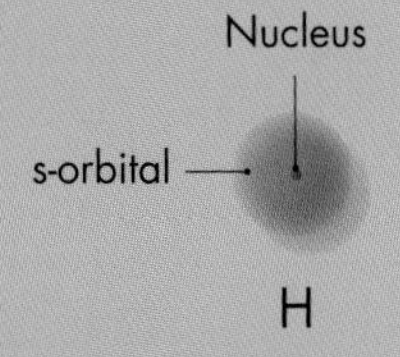

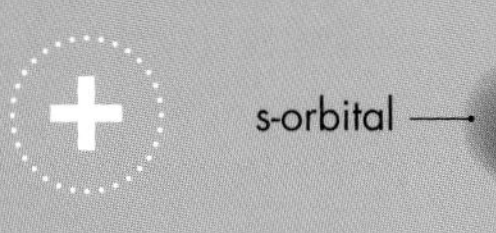

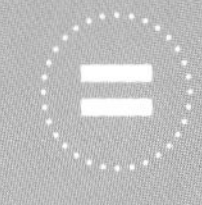

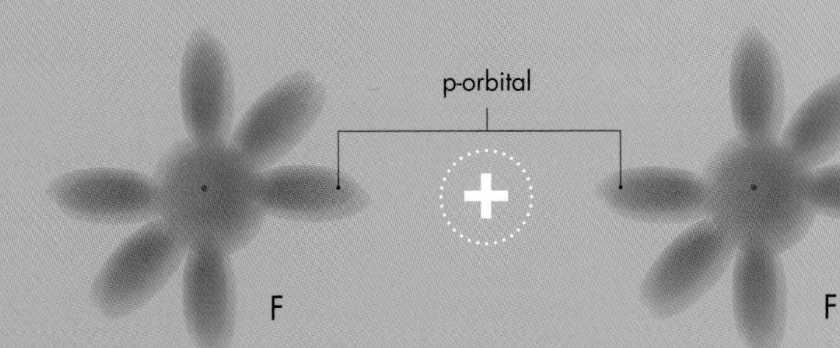

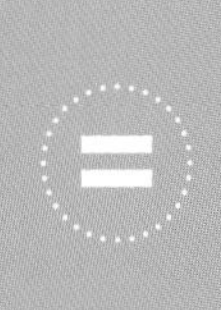

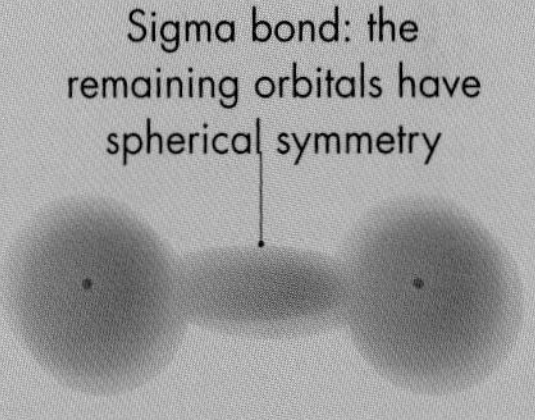

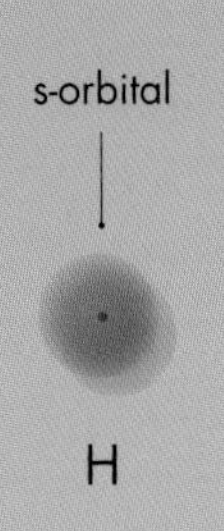

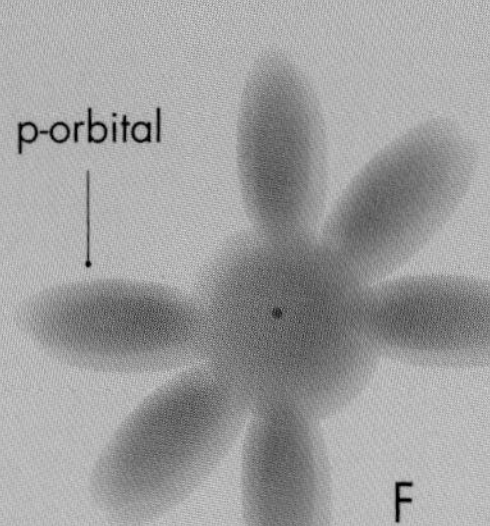

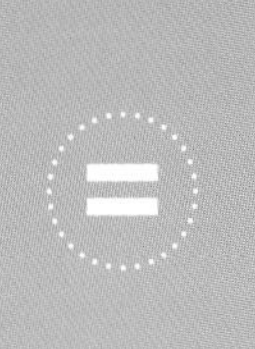

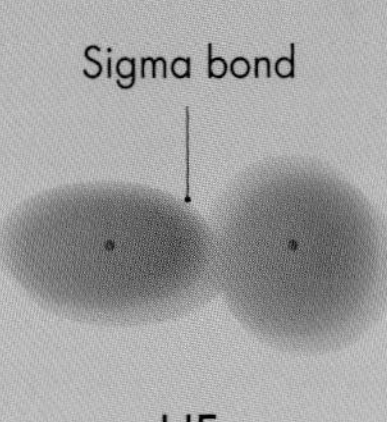

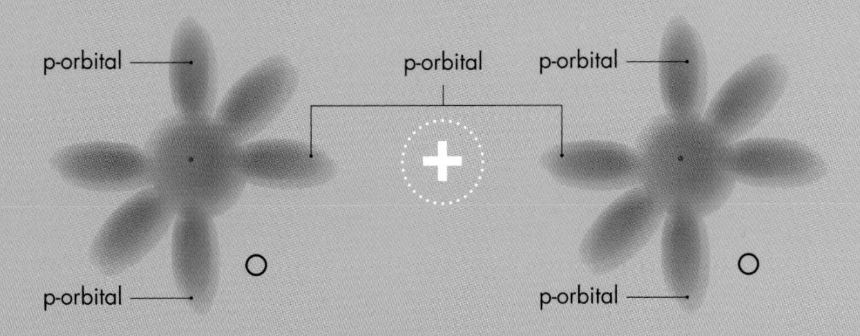

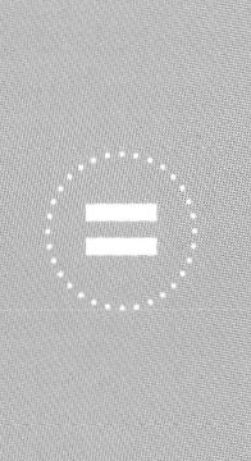

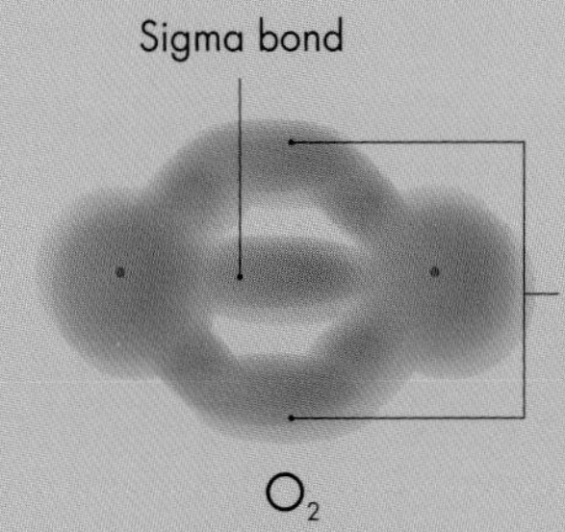

HYDROGEN BONDING

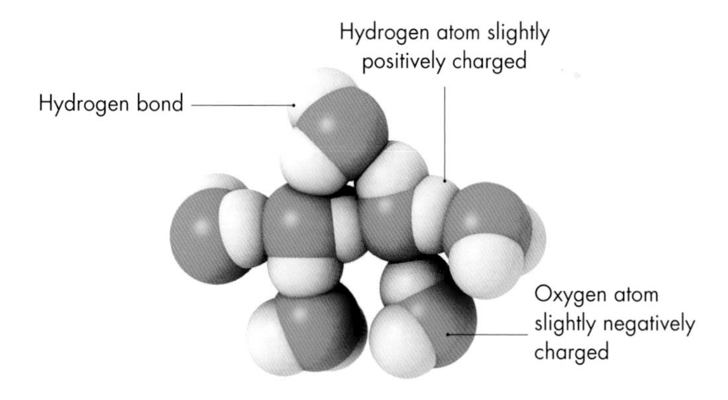

The polarity of water molecules makes them cling together more than they would otherwise do, giving water higher melting and boiling points than it would otherwise have.

WATER: A GOOD SOLVENT

The polarity of water molecules, with a slight negative charge around the oxygen atom and a slight positive charge around the hydrogen atoms, explains why water is so good at dissolving ionic solids. Here, water molecules cluster around sodium ions (purple) and chloride ions (green), with their charged parts attracted to the opposite charges of the ions.

The polarity of water molecules is important in other ways, too. For example, it creates a mutual attraction between two or more water molecules: an intermolecular bond. All kinds of molecules have a slight attraction to each other, the result of small variations in the distribution of electric charge in their bonds. But in water, and in other hydrogen-bearing molecules, the intermolecular attraction is fairly strong, because of the polar nature of the bonds. The intermolecular attraction caused by the presence of hydrogen in a molecule is called a hydrogen bond. In water, the positive region around a hydrogen atom in one molecule clings to the negative region around the oxygen atoms of another—and is the reason why water has higher melting and boiling points than it would otherwise have.

Hydrogen bonding also explains water's strong surface tension, which pulls water into tight, spherical droplets. Molecules at the surface are pulled inward by their mutual attraction to molecules in the bulk of the droplet.

Hydrogen bonding also holds the two strands of a DNA molecule together, but not so strongly that they cannot be unzipped when a section of the DNA is to be copied. Hydrogen atoms at certain points in large protein molecules pull those molecules into particular shapes. Because the shape of a protein molecule is often crucial to its function, and the unzipping of DNA is clearly vital, too, it is no overstatement to say that all life on Earth depends upon hydrogen bonding.

The mutual attraction of water molecules caused by hydrogen bonding explains why water forms such tight droplets, as seen on this leaf.

CARBON: VERSATILE BONDING

Life on Earth also depends, of course, on the element carbon, whose atoms are the building blocks of organic molecules, such as sugars and fats, as well as proteins and DNA. Carbon atoms really are masters of covalent bonding. They readily form single (sigma) bonds with hydrogen atoms and with the atoms of many other elements. They also readily form double bonds with oxygen, sulfur, or nitrogen atoms, as well as single, double, and triple bonds with themselves. Carbon atoms even form rings, composed of six carbon atoms sharing all their valence electrons in a circular orbital. A single polymer molecule can consist of tens of thousands of carbon atoms bonded with atoms of other elements.

Diamond is a form of pure carbon, and any piece of diamond may be considered as a single molecule. The carbon atoms are bonded together in a repeating tetrahedral pattern that gives the diamond its strength. The tetrahedral arrangement of carbon atoms in diamond is due to a phenomenon known as sp^3 hybridization. Carbon's valence shell has eight available slots for electrons to occupy: two in an s-orbital, and a total of six in three p-orbitals. But carbon has only four electrons in its valence shell—and instead of having half-filled s- and p-orbitals, a carbon atom readily forms four equivalent "hybrid" orbitals. The new orbitals are all equally spaced in three dimensions, and so form a perfect tetrahedron. As well as diamond, many much smaller molecules exhibit tetrahedral bonds as a result of carbon's sp^3 hybrid orbitals. Methane (CH_4) is a good example. Carbon can also form other types of hybrid orbital, as well as forming bonds with its "standard" set of s- and p-orbitals. It is this flexibility of bonding that makes carbon such a versatile and important element. Seen here is a selection of carbon-based (organic) molecules that show the versatility of carbon atoms in forming bonds with other carbon atoms and with atoms of other elements.

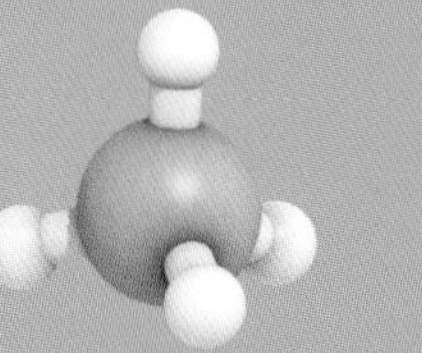

Methane is a tetrahedral molecule, with a hydrogen atom held with sigma bonds, one at each of the sp3 hybrid orbitals

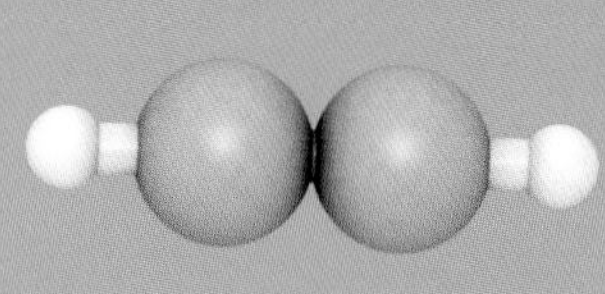

Ethyne is a hydrocarbon in which a triple bond (a sigma bond and two pi bonds) hold the carbon atoms together

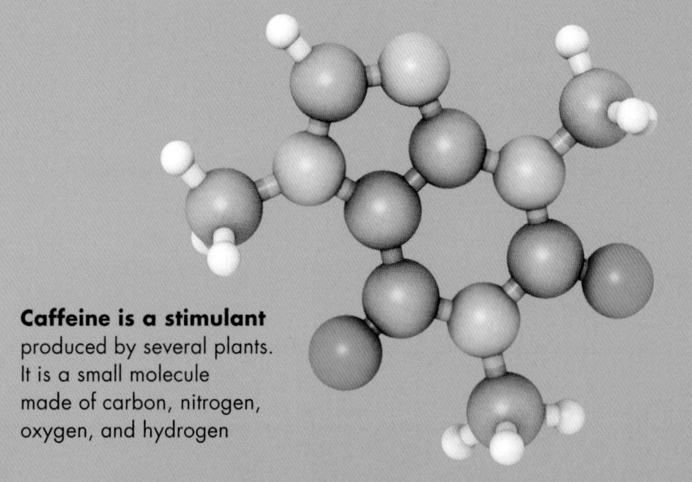

Caffeine is a stimulant produced by several plants. It is a small molecule made of carbon, nitrogen, oxygen, and hydrogen

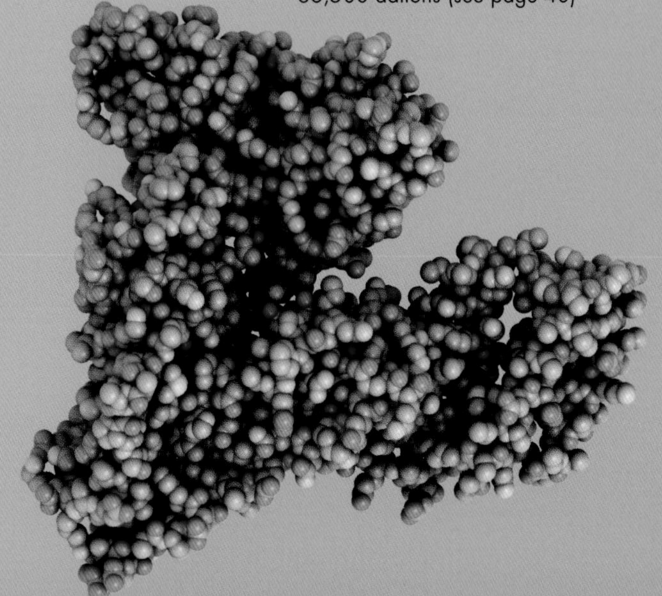

Serum albumin is a protein found in blood. Human serum albumin has a total mass of around 66,500 daltons (see page 40)

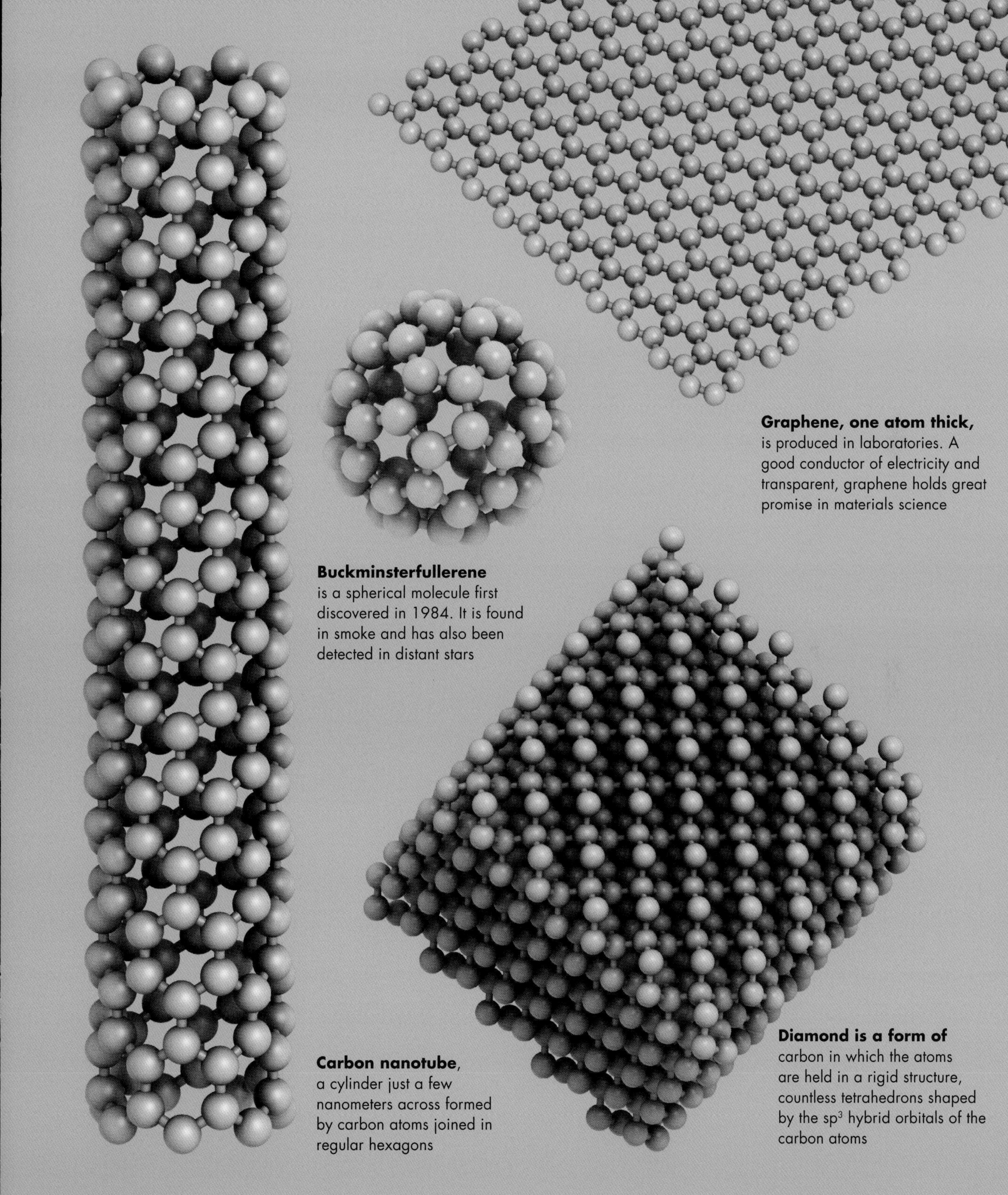

Graphene, one atom thick, is produced in laboratories. A good conductor of electricity and transparent, graphene holds great promise in materials science

Buckminsterfullerene is a spherical molecule first discovered in 1984. It is found in smoke and has also been detected in distant stars

Carbon nanotube, a cylinder just a few nanometers across formed by carbon atoms joined in regular hexagons

Diamond is a form of carbon in which the atoms are held in a rigid structure, countless tetrahedrons shaped by the sp^3 hybrid orbitals of the carbon atoms

MOLECULES AND LIGHT

Diamond is transparent: light passes through it unimpeded. Light is an electromagnetic wave (and at the same time, of course, a stream of photons, see page 33), so it can interact with electrons. One form of pure carbon, buckminsterfullerene, is black, because the electrons forming the bonds are free enough to absorb light. But in diamond, all the electrons are committed to their sp^3 hybrid orbitals; they are engaged in bonding. That is why light passes through. Other forms of electromagnetic radiation beyond the visible spectrum, such as infrared and ultraviolet, can also interact, or not, with electrons. In some cases, the radiation will cause a molecule to twist or turn. In others, it will promote an electron to a higher- energy level within the molecule, or even break the molecular bonds altogether.

A polar molecule, such as water, will rotate back and forth when exposed to electromagnetic radiation, although to varying amounts, depending upon the frequency of the radiation. So water is transparent to light, for example, but it does absorb infrared and microwave radiation. Water vapor and water droplets in the atmosphere absorb some of the infrared radiation emitted by the warm surface of Earth, which would otherwise carry energy out to space, away from the planet. As the water molecules absorb energy from the infrared radiation, the temperature of the atmosphere increases and the water molecules radiate in all directions. Some of that radiation makes it out to space but, importantly, some radiates back down to Earth. This is the process by which our planet's surface is maintained at a higher temperature

than it would be with no atmosphere: the "greenhouse effect." Other molecules—notably carbon dioxide and methane—are also key players in the greenhouse effect. The higher the concentration of these greenhouse gases in the atmosphere, the higher the average surface temperature of Earth becomes.

The interaction between electromagnetic radiation and molecules can create a characteristic spectrum, similar to the way elements can be identified by spectroscopy (see page 68). Molecular spectroscopy has made it possible to ascertain the identity of hundreds of compounds in nebulas deep in space, including some of the compounds that, on Earth, are involved in living processes. When X-rays or gamma rays hit DNA molecules, they can break bonds between the atoms, introducing errors into the genetic code carried by the DNA molecule. These errors are mutations, which can lead to cancer, but they are also an important driving force in evolution.

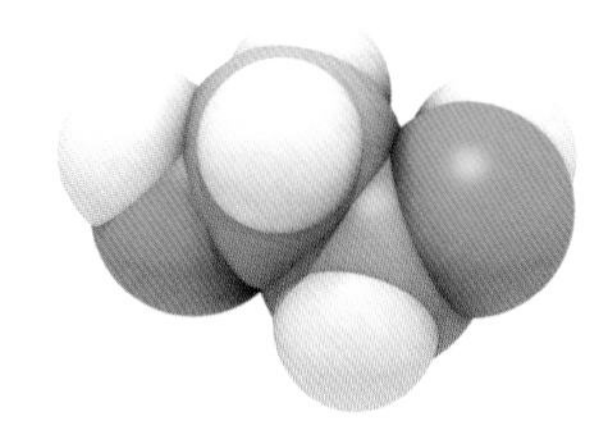

Ethylene glycol, which is the main constituent in automotive antifreeze, was detected in 2002 in a distant cloud of gas near the center of our galaxy, using molecular spectroscopy.

Multiwavelength view, provided by the APEX Telescope, in Chile, of the center of our galaxy (left), where many interesting molecules have been detected, including ethylene glycol (see above).

Molecular spectroscopy, carried out by the Spitzer Space Telescope, detected water and a range of hydrocarbon molecules in a galaxy 3.2 billion light-years away (below). The graph (right) shows how different molecules absorb specific wavelengths across the infrared spectrum.

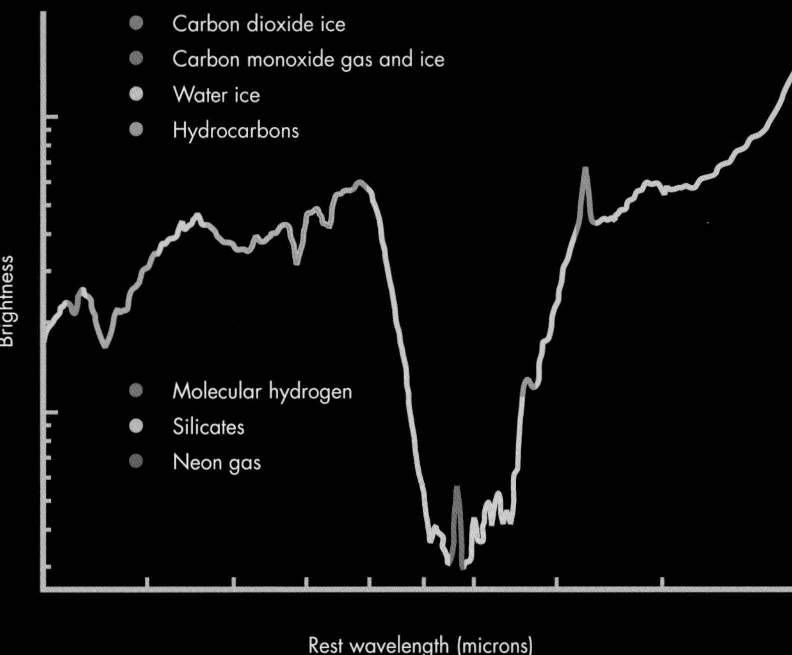

The fact that many molecules only absorb some frequencies of electromagnetic radiation is also the explanation for the different colors of pigments. Chlorophyll in plants, for example, absorbs the ultraviolet, blue, and red parts of the visible spectrum; it uses the energy it receives from these to drive photosynthesis. The other parts of the spectrum pass through unimpeded or are reflected, giving plants their characteristic green color.

The electrons in diamond are locked in their bonds, making them unable to interact with electromagnetic radiation. But there are many substances in which the electrons are free to do so. In a metal, for example, the electrons are so available that they can absorb almost all frequencies of electromagnetic radiation. Typically, the radiation is absorbed, then immediately reradiated in the opposite direction. This is why metals are reflective. The reason why the electrons are so available in metals has to do with the way in which metal atoms bond.

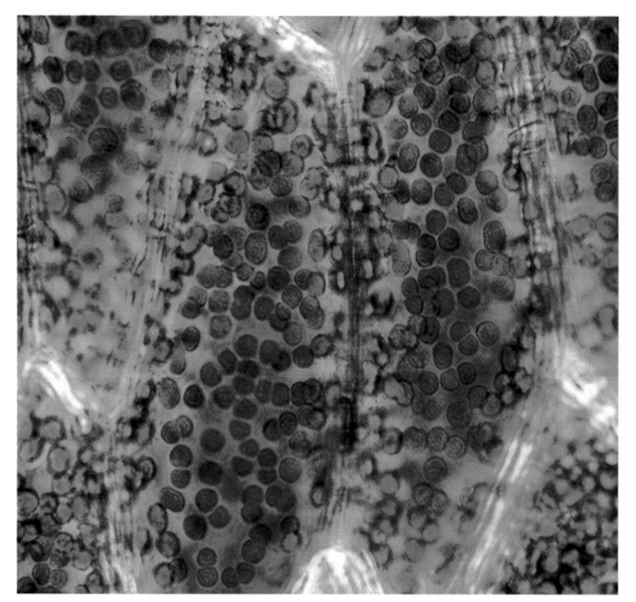

METALLIC BONDING

Put a few trillion iron atoms together (at room temperature) and they will cling tightly to one other and form a solid. They are not held together by covalent bonds, nor by ionic bonds. Instead, they are held in place within a vast (on the atomic scale) sea of shared electrons. The energy levels of the valence electrons of a metal atom are very close together. When you put many metal atoms together, the energy levels merge into one "band" of energy, and the electrons become "delocalized" from their host atoms, free to move around. The free movement of electrons explains why metals are such good conductors of electricity—and is also the reason why metals are such good conductors of heat (the electrons are good at passing on vibrations). It is not surprising, then, that the continuous band of energies shared between the metal atoms is called the conduction band.

Nonmetal atoms do not form a conduction band and tend to be poor conductors of heat and electricity, nor do they reflect electromagnetic radiation in the same way that metals do. There is a class of elements (and compounds) whose conductivity can vary depending upon temperature, electric and magnetic fields, or the presence of other elements. These "semiconductors" are the basis of digital technology, and they will be explored in chapter six.

This photograph, taken with polarized light through a microscope, shows a close-up view of a leaf of *Hookeria* moss. The leaf of this plant is just one cell thick, so we can see the cell wall (blue) and the chloroplasts (green). It is in the chloroplasts that chlorophyll absorbs photons from sunlight and uses the energy of those photons to power the process of photosynthesis.

METALLIC BONDING

The atoms in a solid metal are held in their positions in a "sea" of electrons that all occupy one mutual and continuous band of energies, called the conduction band. Because each electron is not attached to one particular atom, it is free to move through the crystal. When an electric field is applied across a metal, electrons move.

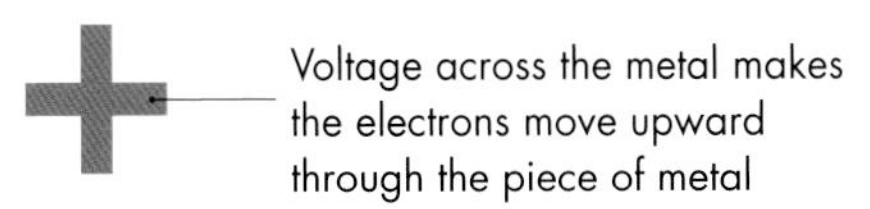

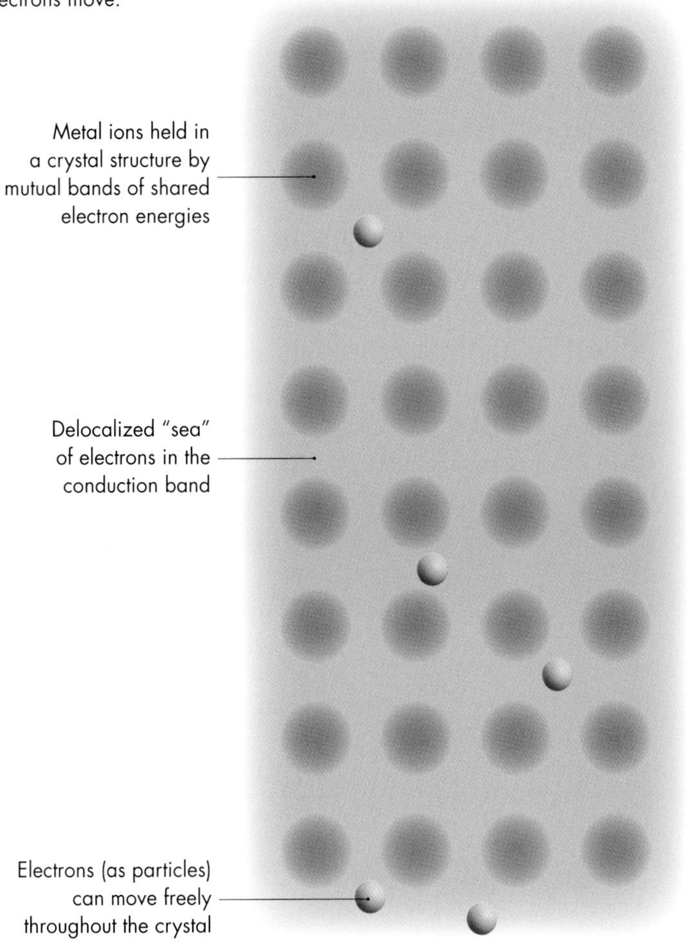

In an isolated atom, the electrons' energy levels are well defined

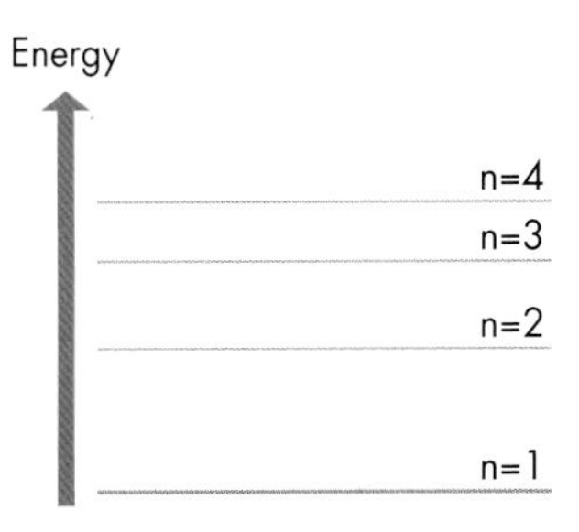

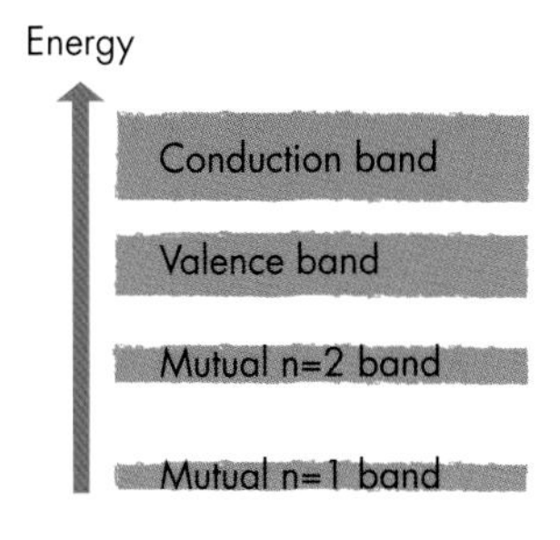

Energy levels change slightly in the presence of many atoms, and so merge into "bands."

CHEMICAL REACTIONS

In any chemical reaction, such as the burning of a candle or the rusting of iron, the compounds or elements present at the end of the reaction are different from those that were present at the beginning. The reason for this is that every chemical reaction involves the breaking and making of interatomic bonds.

BOND ENERGY

Two hydrogen atoms bond together because it is favorable in terms of energy. They have lower energy when bound together than they do when they are separate from each other. It is possible to break the bond, but it takes energy to do so. The amount of energy needed to break a bond is called the bond energy. A chemical reaction only can start if there is enough energy available to break existing bonds. Forming a new bond releases an amount of energy equal to that bond's bond energy.

The energy required to break a bond, and start a reaction, is called the activation energy. In reactions involving gases, the activation energy is provided when gas particles collide with other gas particles, or with the particles of a solid or liquid. Leave pure iron out in the air, for example, and oxygen molecules dashing around in the air will hit the iron surface. Some will provide enough energy to break the bonds between the two oxygen atoms and to interrupt the metallic bonding of some of the iron atoms in the surface of the metal. The iron atoms can then bond to oxygen atoms, forming iron oxide. Many other reactions are also possible at the iron-air interface, involving water molecules and other compounds in the air, such as sulfur dioxide. The result is a complex mixture of compounds on the surface of the iron that we call rust.

Rust forms on iron gradually, as oxygen atoms bond with iron atoms. The oxygen atoms and iron atoms that take part in the reaction become available when oxygen molecules hit iron atoms with enough energy to break the molecules apart and to release iron atoms from the iron crystal.

In the heat of a candle flame, molecules of oxygen and the hydrocarbon molecules of the wax break apart, making their atoms available to form new bonds. The products of the reaction are carbon dioxide (CO_2) and water (H_2O).

A wax candle is a huge tangle of hydrocarbon molecules. These molecules are only made of carbon and hydrogen atoms. A flame held to the wick provides enough energy to melt the wax, which climbs up the wick by capillary action (the movement of a liquid through small spaces in a material due to an attractive force between the liquid and the material). Some of the wax will vaporize in the heat of the flame, and the hydrocarbon molecules can then collide with oxygen molecules in the air. There is enough energy to break the bonds between the oxygen atoms in the oxygen molecules, and the carbon-hydrogen bonds in the hydrocarbon molecules. The result is a transition state in which atoms are free and can form new bonds. Carbon atoms join with oxygen atoms to form carbon dioxide, and hydrogen atoms bond with oxygen atoms to form water. The total bond energy in the water and carbon dioxide molecules ("the products") is less than the total bond energy in the hydrocarbon and oxygen molecules ("the reactants"). The excess energy is released as heat when the new bonds form. The heat sustains the reaction so the candle continues to burn. A reaction, such as the combustion of wax, in which the bond energy of the final products is less than the bond energy of the reactants, is described as exothermic—because it releases energy—often, but not always—as heat. Most reactions are exothermic, because they are energetically favored. But some reactions are endothermic reactions, in which the products have more energy than the reactants. You might expect an endothermic reaction to happen in reverse immediately: for the products to "fall back" to their lower-energy state and form the lower-energy bonds that were present before the reaction. However, they can find a new stable state, albeit with higher bond energies.

ENERGY IN CHEMICAL REACTIONS

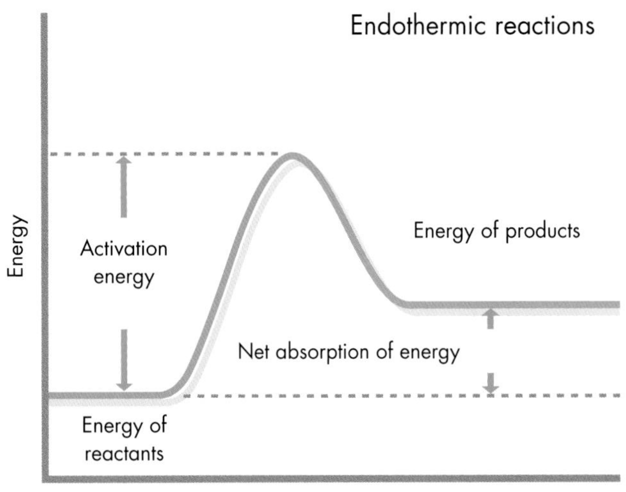

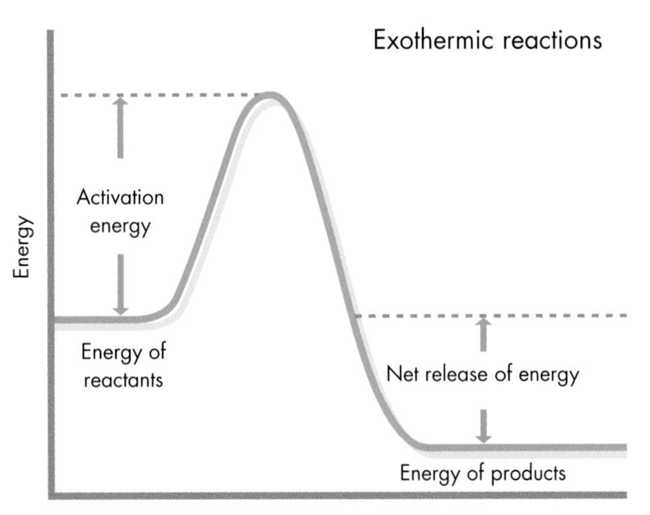

Every chemical reaction requires an input of energy to initiate it. In endothermic reactions (top graph), the energy in the bonds present after the reaction is greater than the energy in the bonds present before. In exothermic reactions (bottom graph), the opposite is true.

To see how this is possible, imagine pushing a car up a hill. This is a useful analogy, especially because graphs of the total energy before and after a chemical reaction really do look like hills. Energy is needed to push the car up the hill, and that energy would be released if you let go of the car on the way up. But if there is a small dip at the top of the hill, and you push the car into that, then it will not roll back down. It will have more energy than it did at the bottom of the hill—the energy you contributed. So it is with endothermic reactions. Supply the necessary activation energy to break the bonds (push the car out of the dip), in the presence of the right reactants, and a new reaction will occur. The making of bonds always releases energy (the bond energy), and that is how the new stable state is possible—represented by the "dip" at the top of the hill.

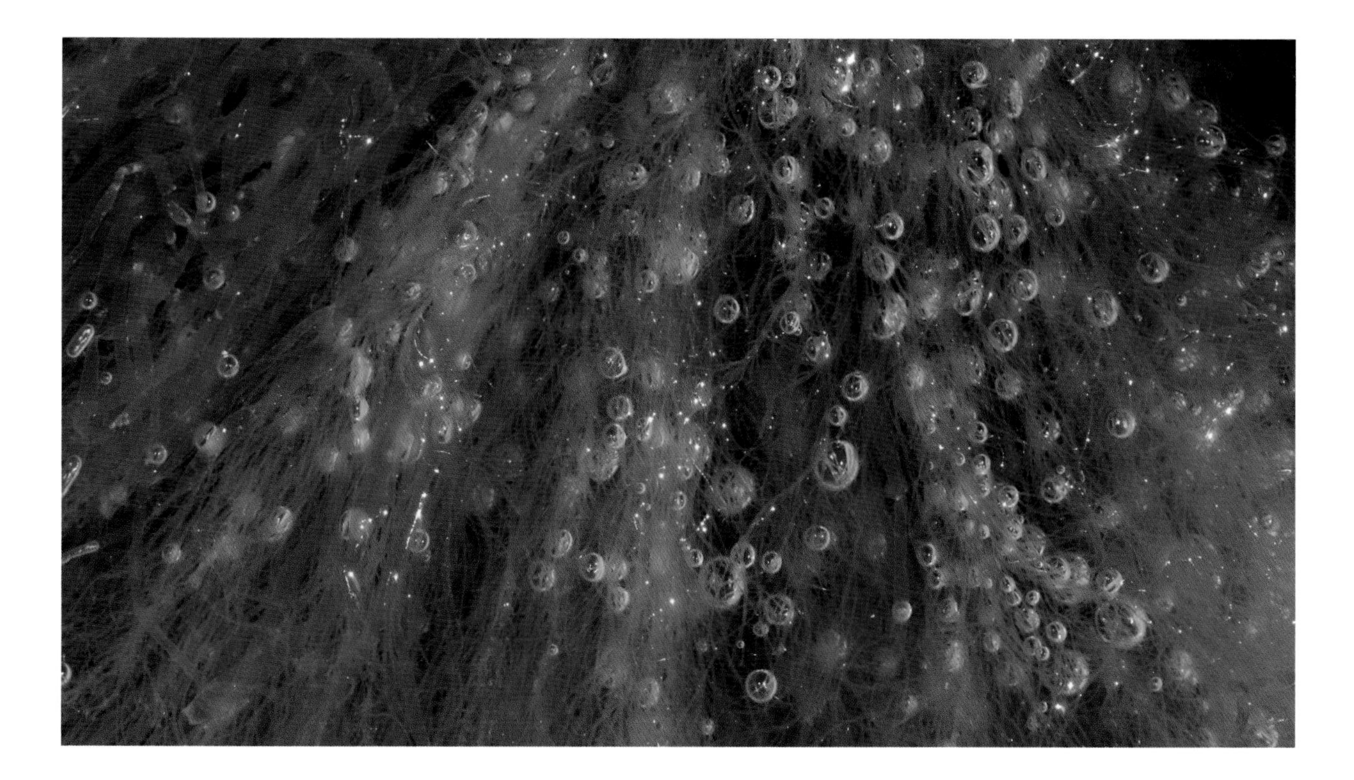

Pondweed produces bubbles of oxygen, which is a waste product of photosynthesis, an overall endothermic set of reactions powered by the energy of sunlight.

REACTIONS OF LIFE

Chemical reactions make life possible. Life depends completely upon the breaking and making of interatomic bonds. Inside every cell of your body, huge numbers of complex chemical reactions are happening all the time. Most of these reactions—including the building of protein molecules and the copying of DNA—require energy; they are endothermic. The energy required is supplied, ultimately, by sunlight. The Sun's energy drives photosynthesis, an overall endothermic series of reactions involving water and carbon dioxide that takes place in plants and certain bacteria. Photosynthesis provides the energy for (nearly) all living things.

The product of photosynthesis is the compound glucose. The total energy of the interatomic bonds in a glucose molecule is much higher than the bonds in the water and carbon dioxide molecules of which they are formed. The energy thus "stored" in a glucose molecule can be released in another series of reactions involving oxygen. The products of those reactions are water and carbon dioxide, and the energy released can be used to power the endothermic reactions that keep the organism alive.

Because paraffin wax is produced from crude oil, which is made from the remains of long-dead marine organisms, the energy available in the wax of a candle was captured millions of years ago. So when we light a candle, we are accessing the energy of ancient sunlight. The same is true when we burn fuel in our cars and coal- or oil-fired power stations. In any reaction like this—simple or complex, exothermic or endothermic—the key players in breaking and making bonds are the electrons. As we will see in the next two chapters, we can manipulate electrons in a host of different ways; for example, in electronic circuits and in microscopes that produce stunning images of actual atoms.

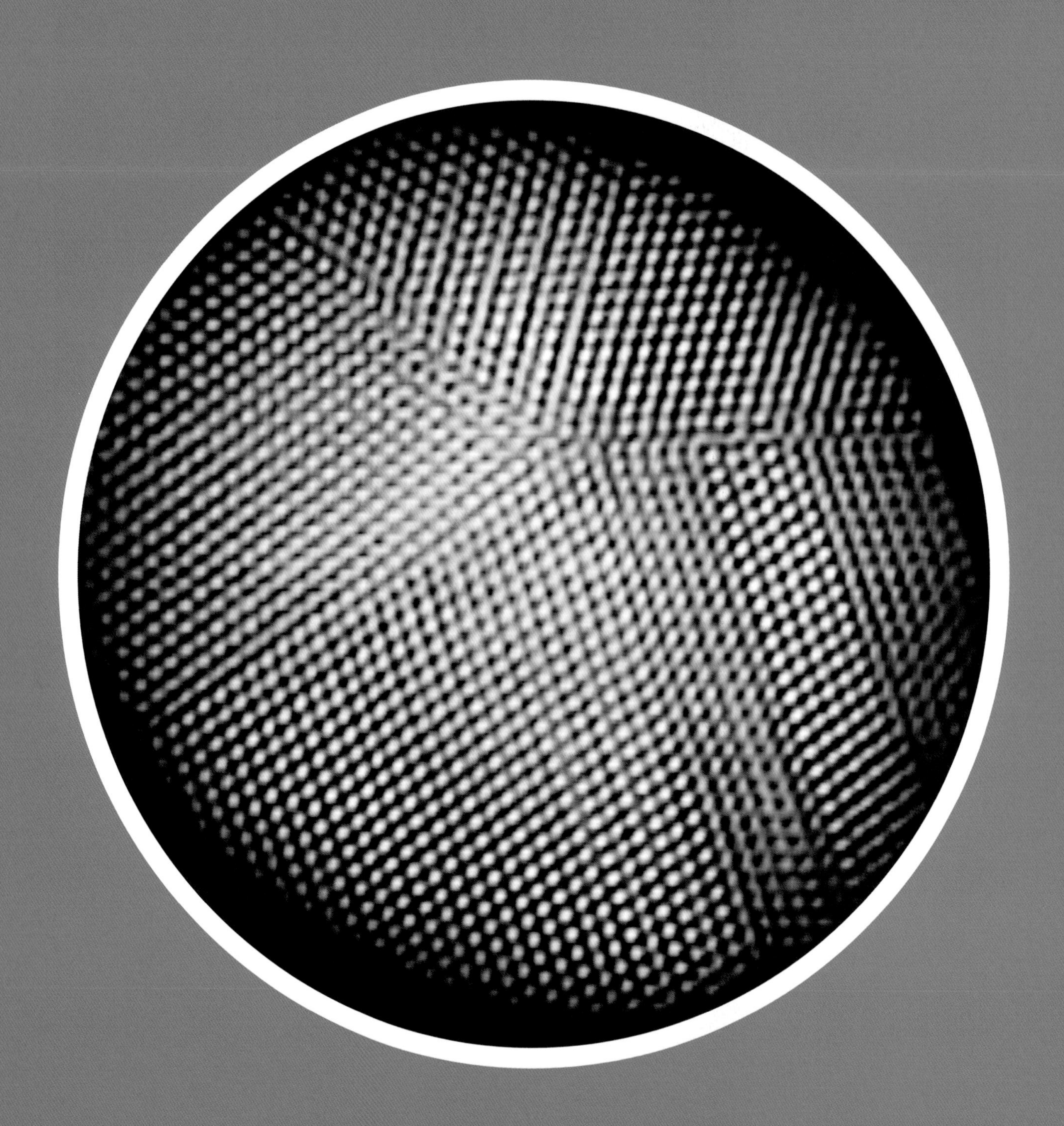

CHAPTER 5

SEEING AND MANIPULATING ATOMS

The difference in scale between atoms and everyday objects is enormous, and the atomic scale seems distant and hard to imagine. But thanks to an array of amazing advances in technology, it is now possible to create images of individual atoms and molecules, and even to manipulate them, one by one. Chemical reactions can be made to happen by directly breaking and making individual bonds between atoms, and the extremely rapid processes that take place in chemical reactions can be observed directly with extremely brief laser pulses.

This remarkable image shows the 27,000 or so atoms that make up a particle of platinum just 2 nanometers in diameter. It was produced by a technique called electron tomography, which involves studying the particle with a transmission electron microscope (see page 116) at a large number of different angles—similar to the way in which a CT (computed tomography) scan can reveal three-dimensional images of organs inside the human body. The image holds information about imperfections in the platinum crystal—information that could help to improve electronic devices, such as light-emitting diodes (LEDs).

WAYS TO SEE ATOMS

The wavelength of light is a few thousand times the size of an atom. This fact makes it impossible to see atoms in the way we see other objects: by bouncing light off them. X-rays have much shorter wavelengths, but most X-rays pass through atoms, and it is not practical to produce an X-ray microscope that can produce images. Electrons, which behave as waves as well as particles, have even smaller wavelengths than X-rays—and they can be used to produce images of individual atoms.

THE CHALLENGE OF RESOLUTION

When you look at a forest from far away, you see a blur of green and brown. You are not able to "resolve" one tree from another:;you cannot see them as two separate objects. If you move closer, you will begin to see the trees. Move closer still, and you are able to resolve individual leaves. If you hold a leaf in your hand, you can see some details on it, but your eyes cannot resolve the individual cells of which the plant is made. The resolution of the human eye is not good enough to do so. The leaf's cells all blur into one, as the forest's trees did from afar. The smallest objects a human eye can resolve are a little less than one-tenth of a millimeter in size (0.004 inch)—about the same size as a human egg cell.

Using a powerful microscope, it becomes possible to see all living cells, and some of the features inside them. In other words, the magnified image that a microscope produces has much greater resolution than the image formed in the naked eye, but there is a limit to the resolution of a light microscope. This limit is not technological but due to the wave nature of light. In the 1870s, German physicist Ernst Abbe discovered that it is not possible to resolve two objects separated by less than about half the wavelength of light. The shortest wavelength visible to the human eye, blue light, has a wavelength of about 4,000 Å; an atom has a diameter of only a few angstroms.

X-rays have a much shorter wavelength—from 100 Å down to 10 Å. This is close to the atomic scale and should certainly be able to resolve large molecules—in theory. However, there is no lens that can focus X-rays in the same way as light, so it is not practical to build an X-ray

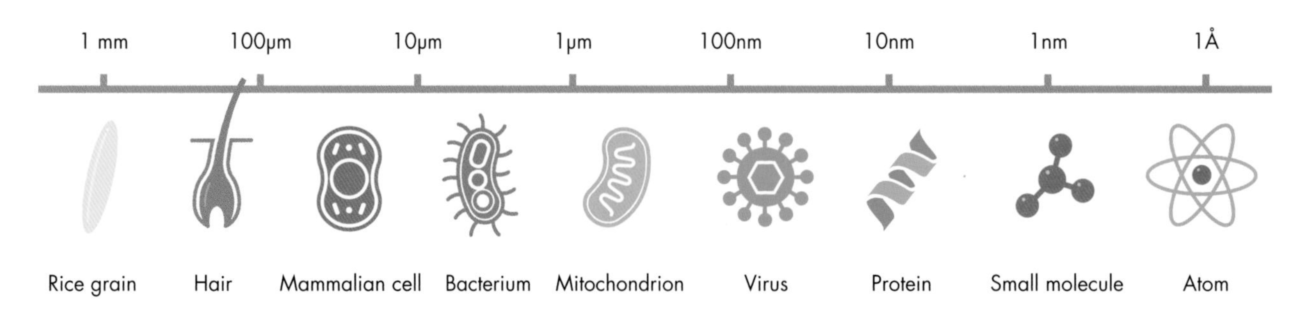

The resolution of the human eye is limited by the distance at which it can clearly focus (a few inches) and how tightly packed the light-sensitive cells are in the retina. The resolution of a light microscope, on the other hand, is limited by the wave nature of light itself. After: © Johan Jarnestad/The Royal Swedish Academy of Sciences

microscope to produce images of molecules. Furthermore, X-rays pass through most atoms instead of bouncing off. However, X-rays do diffract strongly as they pass through a substance. Diffraction is the bending of waves as they move past an edge or pass through a gap (see page 50); it is most noticeable when the gap is of comparable size to the wavelength.

The diffraction of X-rays through a regular array of atoms (or ions), as in a crystal, gives rise to regular patterns on a screen or a photographic plate. Beginning in the 1910s, X-ray crystallography has made it possible to work out the structure of crystals and molecules. But those wanting to produce images of actual atoms would have to wait for new techniques and technological advances.

Pass X-rays through a thin slice of aquamarine crystal (left) and the rays diffract strongly, because their wavelength is comparable to the atoms in the crystal. The diffracted rays interfere (see page 51), producing a pattern of dots that can be recorded on a photographic plate (above). The pattern, different for each compound, holds information about the positions of the atoms in the crystal (see page 96).

USING ELECTRONS

In the 1930s, a new type of microscope became available that uses electrons instead of light. As with all particles, electrons also behave as waves; their wavelengths depend upon their energy, and high-energy electrons can have wavelengths of around 1 picometer (pm): 0.01 Å. This makes them, in principle, easily able to resolve atoms. There are two main types of electron microscope: the scanning electron microscope (SEM) and the transmission electron microscope (TEM).

Transmission electron micrograph of a mitochondrion—a structure in a cell that is a few hundred nanometers across. The folded membranes inside the mitochondrion, each a few nanometers thick, are easily resolved.

Inside an SEM, a beam of electrons scans across a prepared sample and a detector picks up the electrons that scatter off, building up a detailed image. The resolution of the image is limited by the width of the beam and the (electromagnetic) lenses that focus the beam and make it scan, but it can be as low as about 100 Å (10 nm). In a TEM, an electron beam passes through a thin sample, instead of bouncing off it. The resolution of a TEM depends upon the lenses as before, and also on the wavelength of the electrons—and that depends upon their energy, and therefore how fast they are accelerated to form the beam. This arrangement typically leads to a resolution of a few ångstroms. From 1933, when the resolution of TEMs first surpassed the diffraction limit of optical microscopes, these microscopes provided biologists with ever more detailed views of the anatomy of biological cells and materials, giving scientists a better understanding of the structure of materials at the nanoscale.

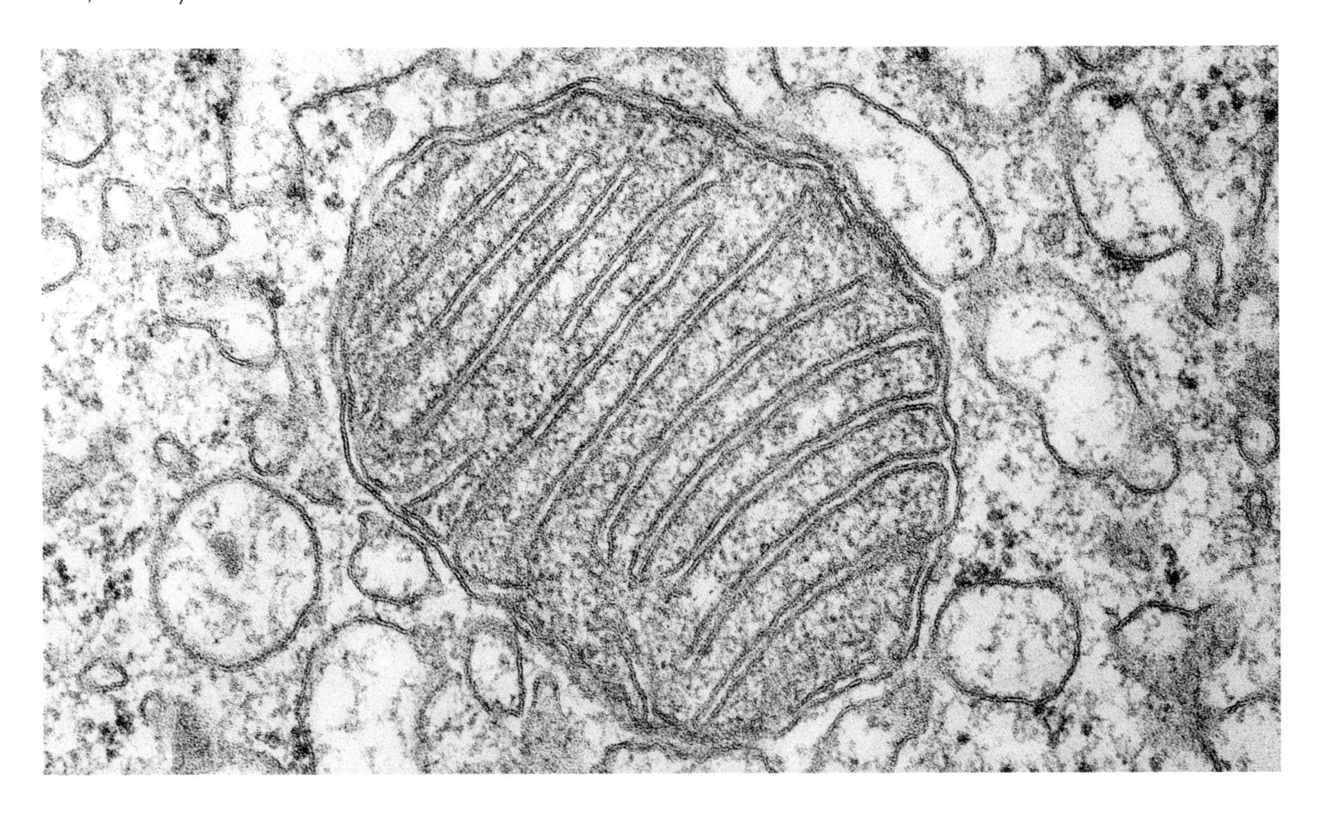

THE FIRST IMAGES OF ATOMS

While electron microscopes were proving invaluable for biologists and materials scientists, they were still not able to produce images of atoms. A very different approach was needed to create the first such images. In 1936, German physicist Erwin Müller invented the field emission microscope (FEM), which produces images with a resolution close to the atomic scale. In an FEM, electrons are pushed away from a very sharp metal tip by a strong electric field. More electrons are released from areas, where the electron density is highest: around the atoms. The electrons follow the straight force lines of the electric field through a high vacuum. They end up on a detector where they form a projected image of the electron density at the surface of the metal tip.

The blurry image of the atoms at the extremely sharp molybdenum tip of a field emission microscope.

In 1951, Müller made a crucial improvement to the device, introducing a small amount of gas, typically helium, into the apparatus. The result was the field ion microscope (FIM) —the first device to produce images of atoms. At very low temperatures, the helium atoms become adsorbed (stuck) to the atoms at the surface. When the electric field is then applied, the helium atoms become ionized and are pushed away from the metal tip. A screen detects the impacts of the helium ions, building up a picture of the atoms at the surface.

Field ion microscopy can produce images only of pure metals—and only those that can be made into extremely sharp tips. In the 1960s, Müller and his colleague John Panitz made a further improvement, allowing (ionized) atoms of the sample itself to fly away from the tip and onto the detector. With this "atom probe," alloys (mixtures of metals) and even compounds can be analyzed. An important feature of the atom probe is a built-in mass spectrometer (see page 69) that measures the mass of the atoms received at the detector and can therefore identify which elements are present. Because the atoms peel away one layer at a time, the atom probe can produce a three-dimensional map of the internal structure of the tip, detailing the positions of atoms of different elements.

The first images of atoms were produced in 1955, by the inventor of the field ion microscope, Erwin Müller. Shown here is an image of the atoms on a platinum tip inside a field ion microscope.

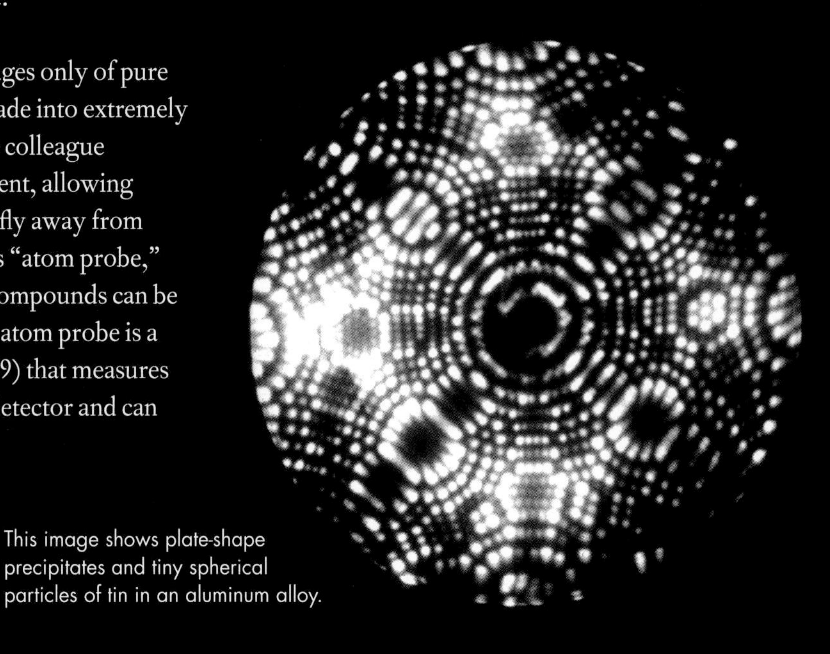

Atom probe microscopy can reveal the positions of atoms, and which element they belong to, in three dimensions.

This image shows plate-shape precipitates and tiny spherical particles of tin in an aluminum alloy.

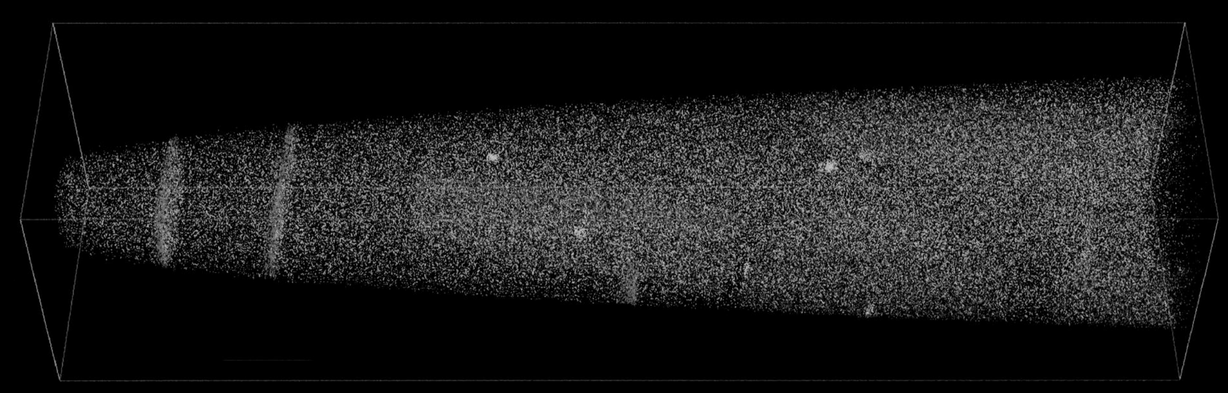

FIELD ION MICROSCOPY

Atoms of an inert gas stick to the atoms at the surface of the sharp tip inside a field ion microscope. A high voltage on the tip ionizes the gas atoms, which then fly out directly to the phosphor screen.

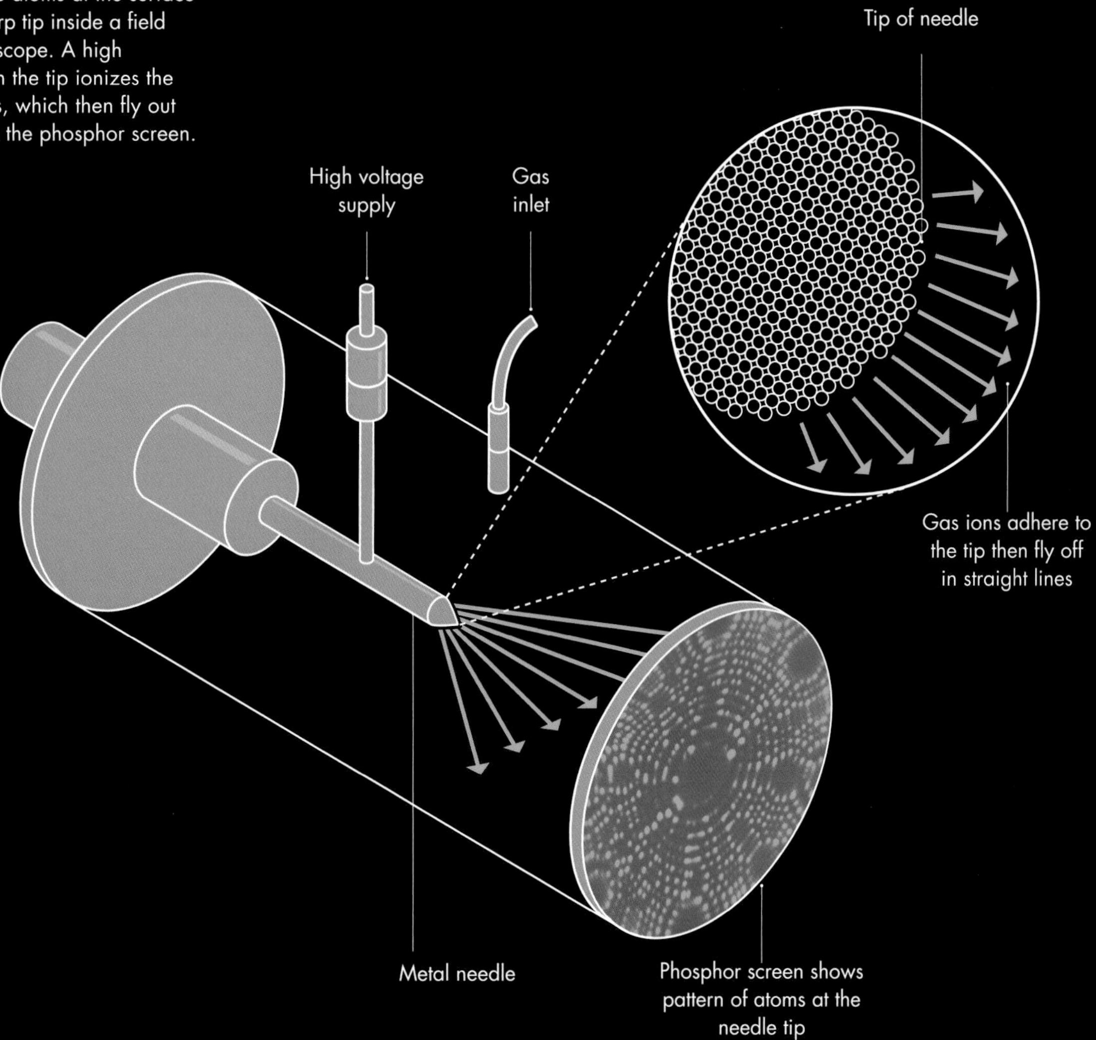

SCANNING TRANSMISSION ELECTRON MICROSCOPE

Adding to the growing array of technologies for producing images at or near atomic resolution—and to the list of acronyms they generate—is STEM: the scanning transmission electron microscope. Like transmission electron microscopy, STEM involves a beam of electrons passing through a thin sample of material. But in STEM, the beam is extremely narrow and is made to scan along successive lines, each one close to the last. This "raster scanning" maximizes the potential resolution of the electron beam and can produce clear images of atoms and atomic bonds. Additionally, detectors collect electrons that scatter off atoms in the sample at high angles, and also X-rays produced when the electron beam hits atoms and interatomic bonds. The energy (and therefore wavelength) of the X-rays depends upon the atomic number of the atom—so different elements can also be differentiated.

The image produced by electrons that have interacted with atoms within the sample and passed through is called the "light field image." The electrons that are scattered to high angles away from the main path of the beam make up the "dark field image." By combining both images and the information from the X-ray detector, STEM can reveal a wealth of information about the sample—and stunning, thought-provoking images.

A selection of false-color STEM images
1 Microcrystal of seven uranium atoms.
2 Graphene (see page 103)—each dot is a carbon atom, and the lines between them are covalent bonds. 3 An island of copper and silver atoms in a "sea" of aluminum atoms.
4 Nanoparticles of iron oxide. 5 A tiny void less than 1 nanometer in diameter cut in a natural diamond (the orange dots are carbon atoms).

1

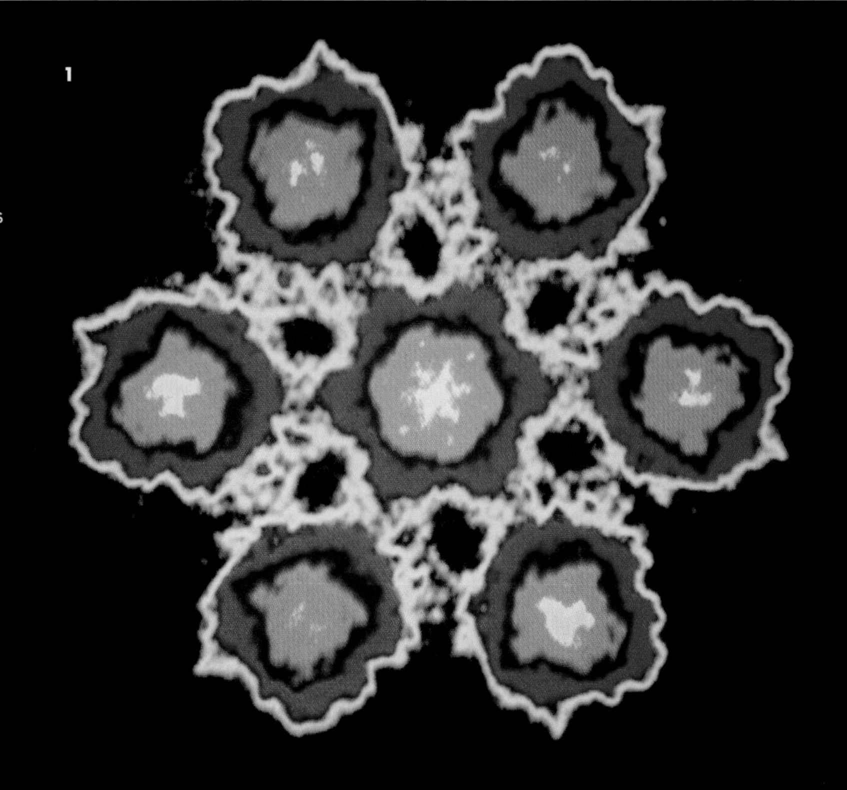

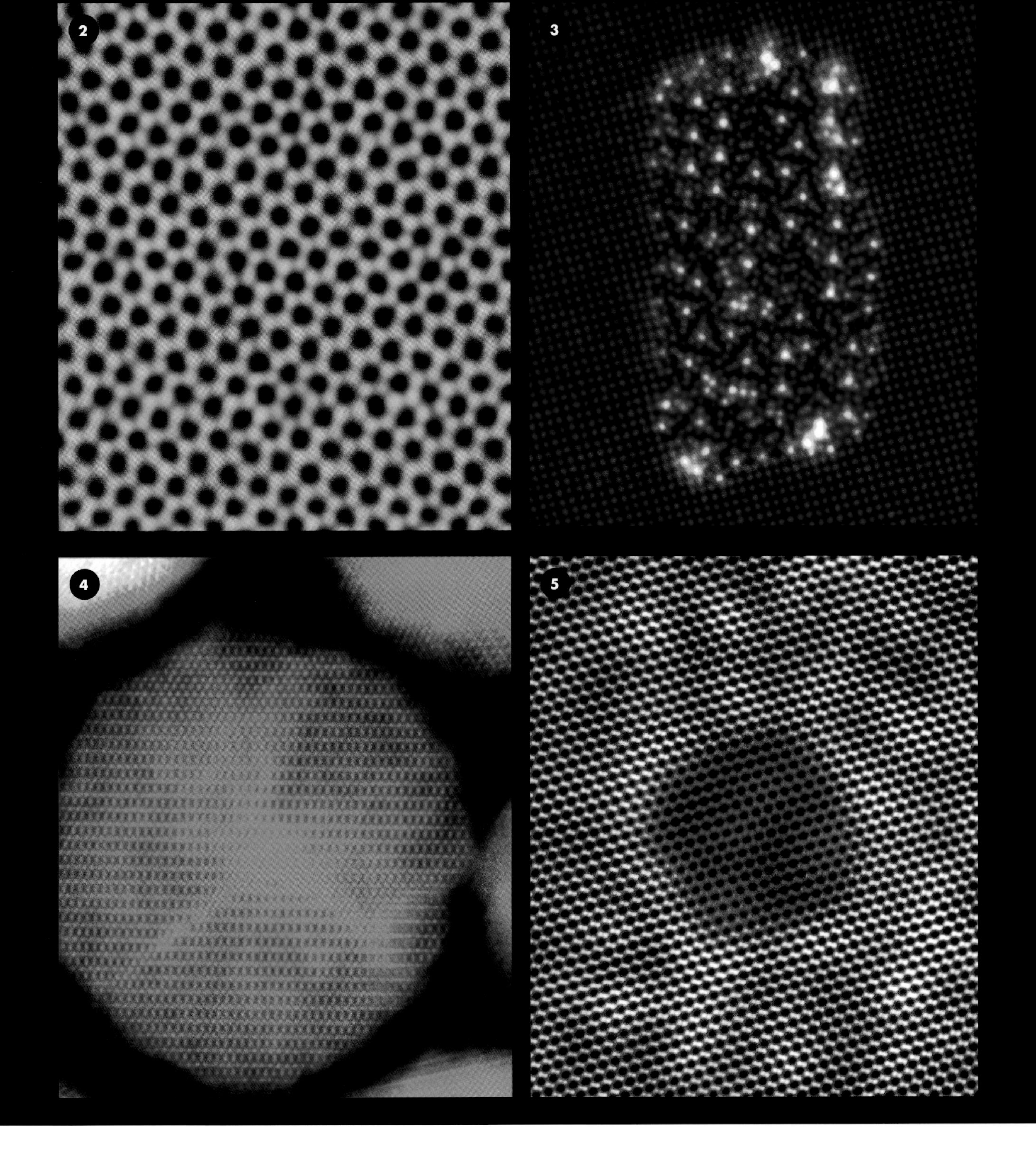
2
3
4
5

SCANNING PROBE MICROSCOPY

The best images of atoms to date have not been produced by X-rays or electron microscopes, nor by field ion microscopy or atom probes. Instead, they are the result of scanning the atomic-scale bumps in the surface of a sample with an extremely sharp probe. In addition to producing incredible images, some scanning probe microscopes can also interact directly with individual atoms and move them around.

SCANNING TUNNELING MICROSCOPE

The year 1981 brought a completely new way of producing faithful images of atoms at a surface, when Swiss physicist Heinrich Rohrer and German physicist Gerd Binnig invented the scanning tunneling microscope (STM). This remarkable device earned its inventors the 1986 Nobel Prize in Physics. The scanning tunneling microscope was the first of a new kind of imaging device: the scanning probe microscope. Instead of capturing an image of the surface directly by shining light, X-rays, or electron beams onto it, a scanning probe microscope builds up an accurate contour map of the surface by detecting the undulations of the surface in a series of scans. A little like reading braille, a scanning probe microscope builds up a picture of the shape of a surface by feeling its bumps.

The main feature of an STM is the probe that scans the surface. This is an extremely sharp metal point, only a few atoms wide at its tip, positioned a few ångstroms above the surface of the sample. The sample itself needs to be

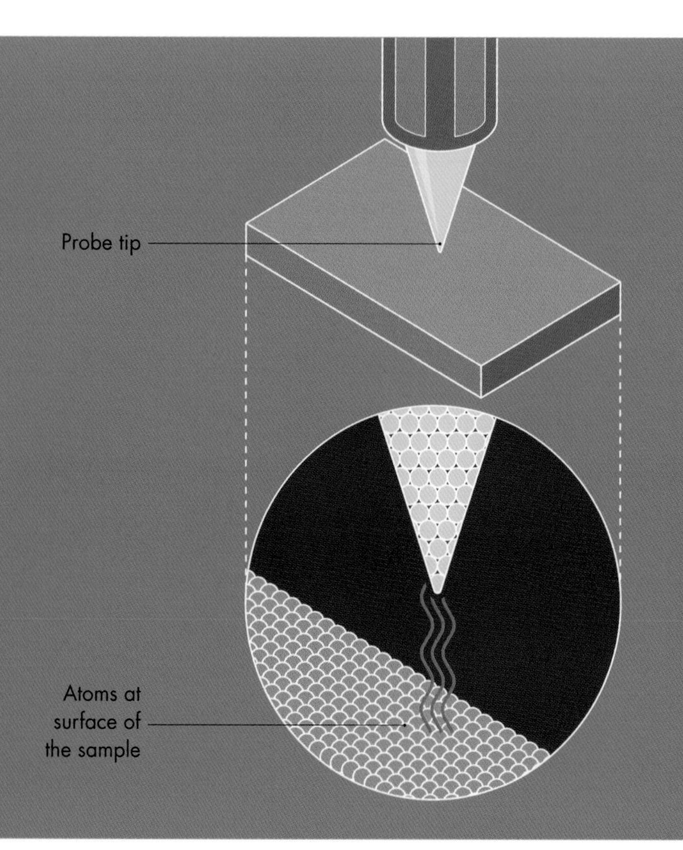

HOW AN STM WORKS

Inside a scanning tunneling microscope, a sharp-tip probe scans in lines along a conducting surface. A small electric current passes between the probe and the atoms at the surface. The current varies according to the distance to the electrons of the atoms at the surface—with enough sensitivity to reveal lumps and bumps of individual atoms.

a conductor, because the process depends upon a small electric current between the tip and the atoms at the surface. The voltage between the probe tip and the surface is not enough to generate a spark between them, even across such a tiny gap. Instead, electrons "leak" across the gap, one by one, by the quantum mechanical phenomenon of tunneling (see page 63). Specifically, the wave functions of the electrons at the surface extend above the surface, and the wave functions of the electrons in the tip extend beyond the tip, so that they overlap slightly (more so at very close separation or when the current is higher). This provides a small probability that electrons can tunnel across the gap. The greater the density of electrons at a particular point in the sample, and the closer the tip is to the surface, the greater that probability becomes—and the greater the tunneling current will be.

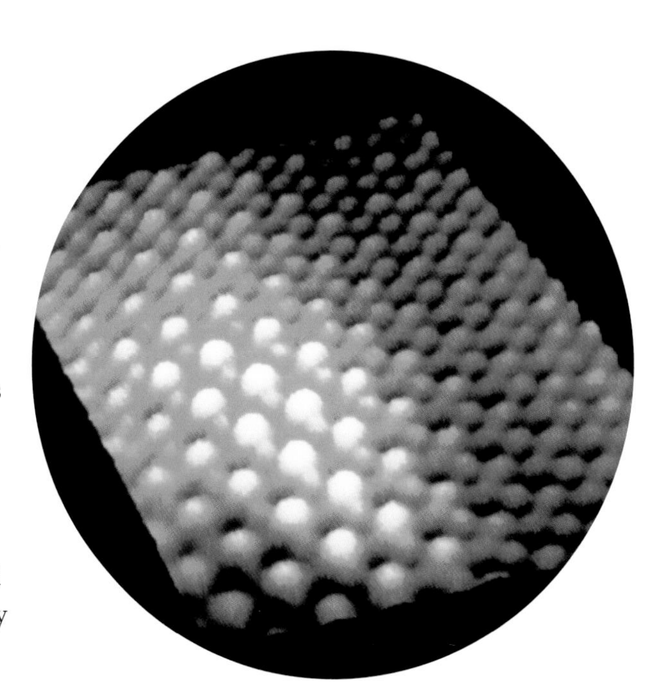

False color STM image of palladium atoms (white) on graphite (carbon atoms, blue). The palladium atoms are about 4 ångstroms apart, the carbon atoms just over 3 ångstroms apart.

Early false color STM image from the 1980s—an intimate view of gold atoms (yellow, red, black) on a graphite surface (carbon atoms, green). The gold atoms have congregated into an island just greater than 1 nanometer wide.

The probe scans along the surface in closely-spaced lines. The device detects the tunneling current and then automatically takes one of two actions. It either changes its height to maintain a constant tunneling current or maintains a constant height and measures the changes in tunneling current. In both cases, with scans along many adjacent lines, the device builds up a very accurate picture of the atom-size bumps in the surface. A computer puts the information together to produce a three-dimensional image.

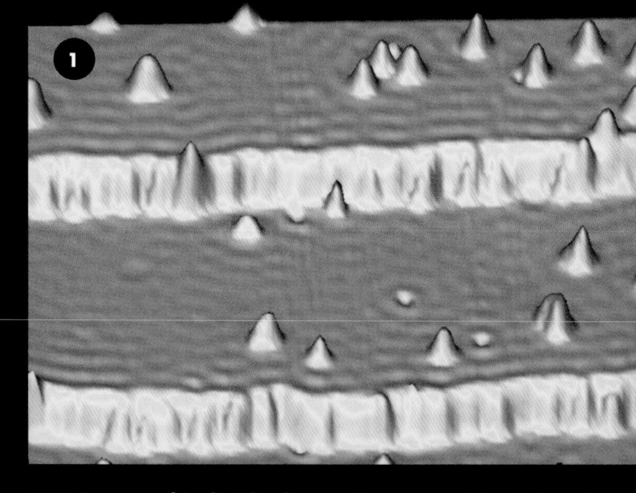

STM image of individual manganese atoms (1) positioned on a gallium arsenide surface, as part of a research project that aims to increase the power and miniaturization of computer chips. False color STM image of sodium atoms (2) (shown here yellow and dark green) and magnesium ions (pink and light green) in a manganese crystal.

3

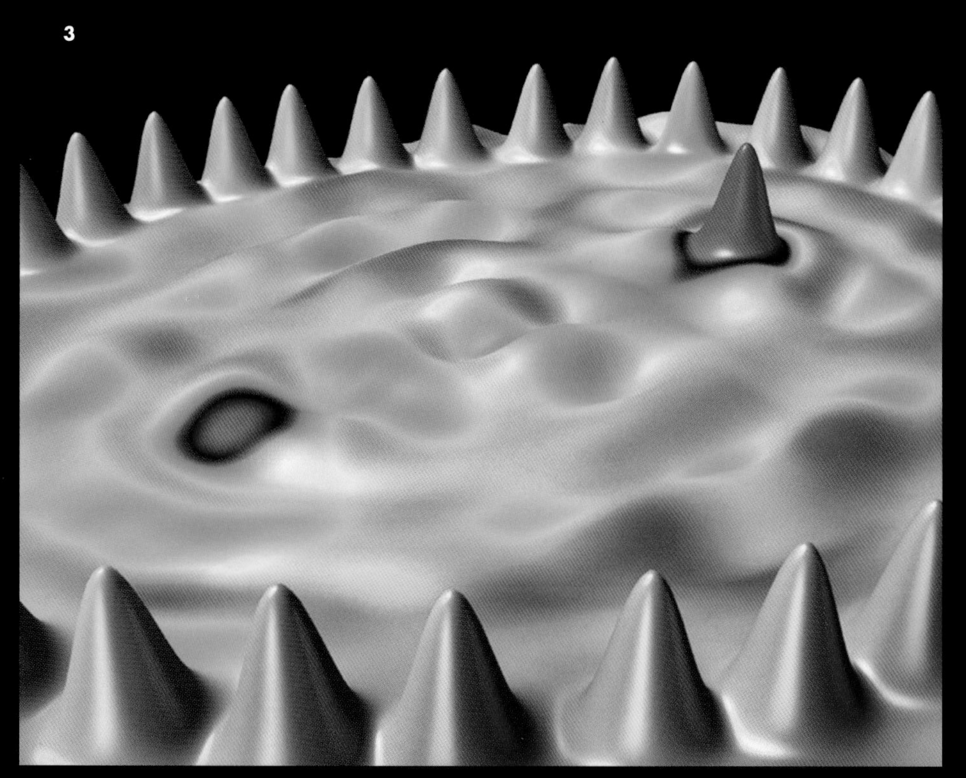

STM image of cobalt atoms (3) arranged in an elliptical structure known as a "corral". Electron waves in the copper substrate interact with a magnetic coral atom (pink), resulting in a mirage effect: a phantom cobalt atom is visible at the other focus of the ellipse. False color STM images showing (4) cobalt atoms (pink circles) on a copper surface, and (5) a stack of carbon nanotubes (see page 103).

ATOMIC FORCE MICROSCOPE

Amazing though it is, the STM has one key limitation: only conductive materials can be imaged, because the tunneling current that is key to its operation must be able to flow. That limitation was overcome in 1986—by the same team who invented the STM—with the invention of the atomic force microscope (AFM). The introduction of the AFM opened up scanning probe microscopy to a whole new array of materials. The principle behind the AFM is very similar to the STM, but instead of measuring a tunneling current, the instrument senses a tiny attractive or repulsive force between probe tip and surface atoms (see box below). The force is attractive at small distances, so the tip is normally positioned far enough away to make the force repulsive, avoiding the problem of the tip "sticking" to the surface.

The probe tip is attached to a flexible cantilever, and because the force varies with the atom-size bumps, the cantilever flexes up and down. The tiny movements of the cantilever are detected by bouncing a laser beam off it and onto a detector. Scanning across the surface in closely-spaced lines as before produces an atomic-scale image on a computer screen. In a variation of AFM, the probe tip is made to oscillate up and down at high frequency, but it does not make contact with the atoms at the surface. In this dynamic, or noncontact, mode, the frequency or the amplitude of the oscillations changes as the surface to tip force varies. With incredibly accurate measurements of the force between the surface atoms and the tip, atomic force microscopy can even identify the elements to which the atoms at the surface belong, atom by atom.

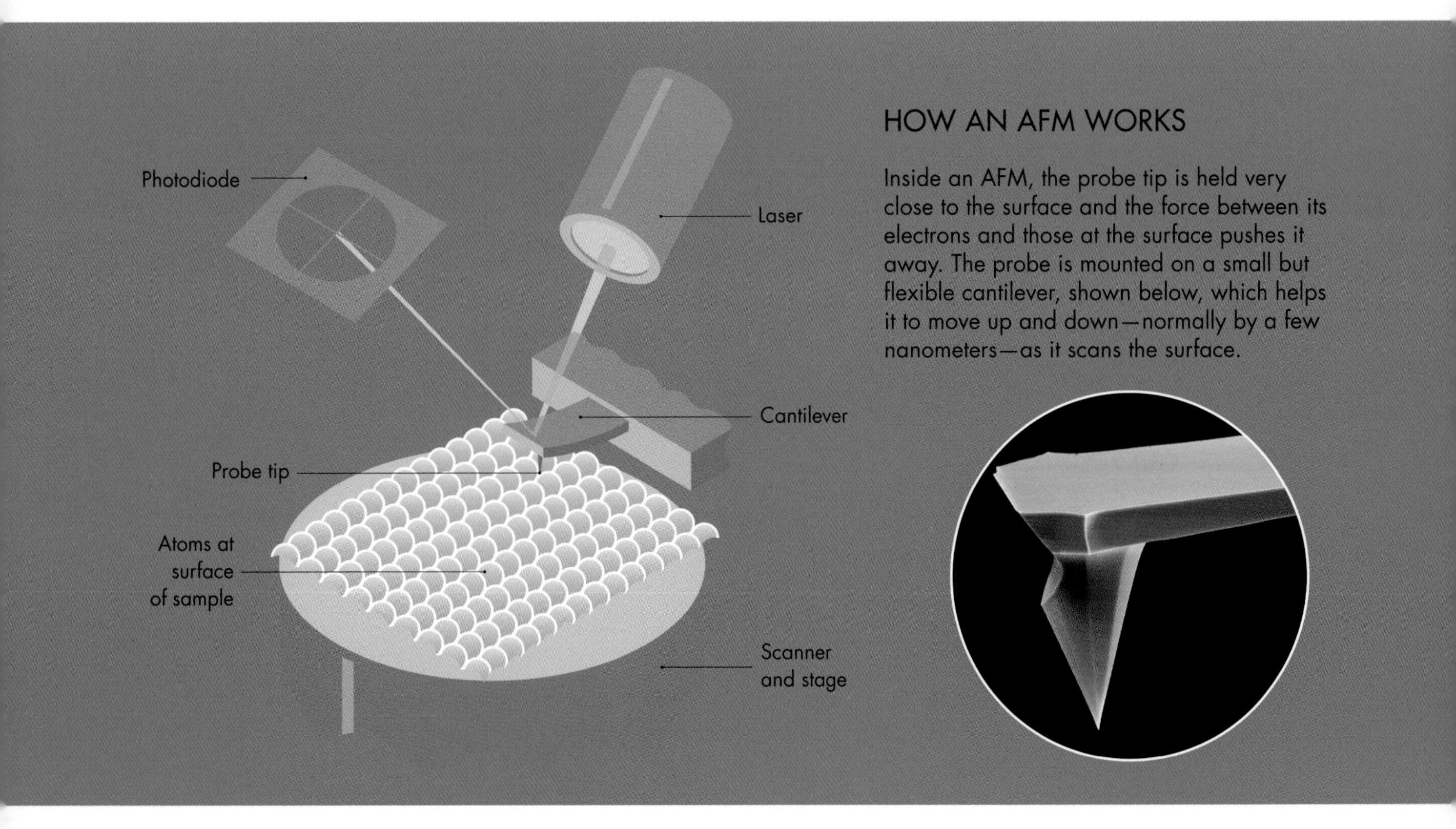

HOW AN AFM WORKS

Inside an AFM, the probe tip is held very close to the surface and the force between its electrons and those at the surface pushes it away. The probe is mounted on a small but flexible cantilever, shown below, which helps it to move up and down—normally by a few nanometers—as it scans the surface.

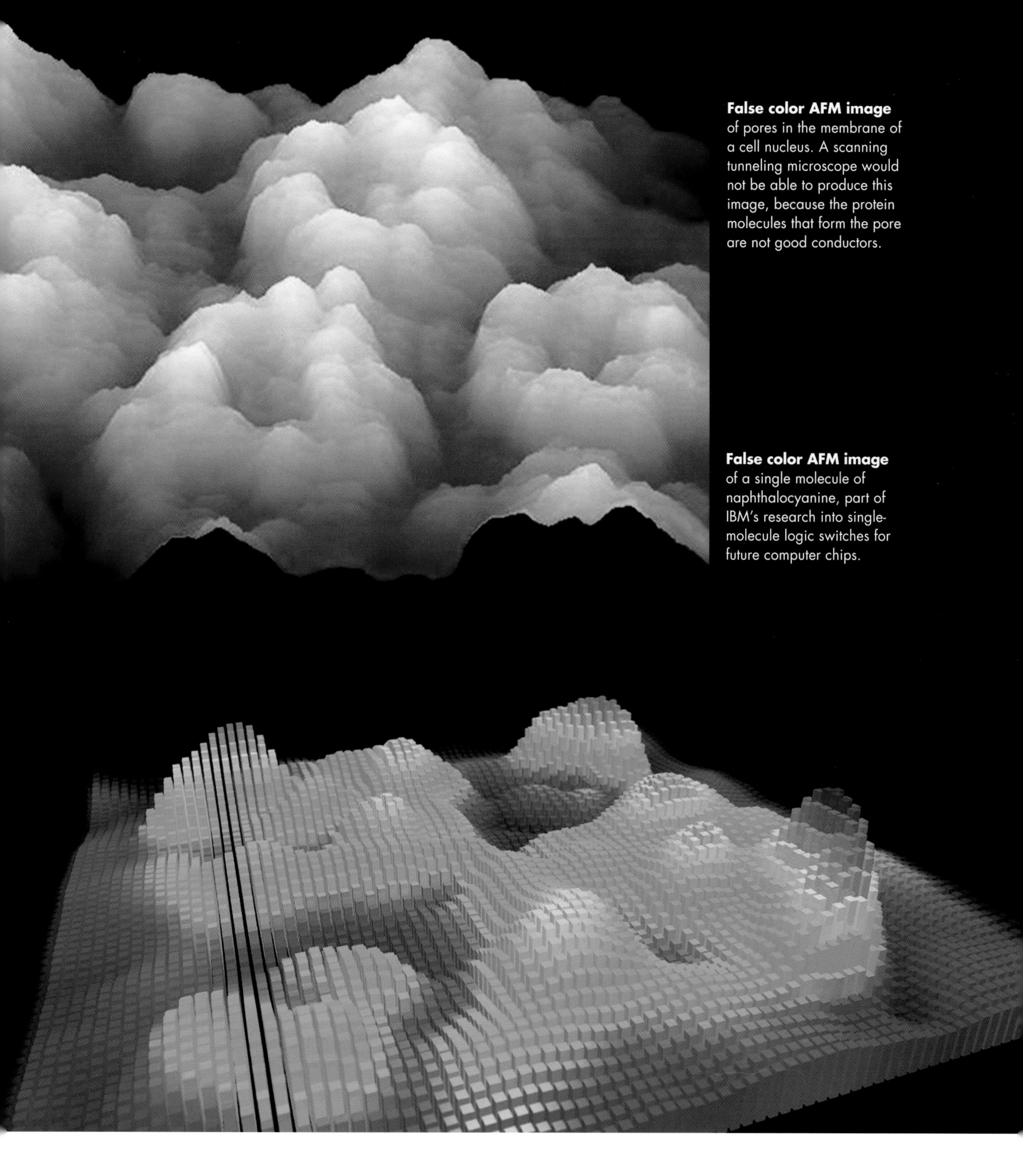

False color AFM image of pores in the membrane of a cell nucleus. A scanning tunneling microscope would not be able to produce this image, because the protein molecules that form the pore are not good conductors.

False color AFM image of a single molecule of naphthalocyanine, part of IBM's research into single-molecule logic switches for future computer chips.

INTERACTING WITH ATOMS

Scanning probe microscopes can do more than produce incredible images of atoms: they can manipulate individual chemical bonds, triggering chemical reactions. Atomic physicists have developed many other ways to interact directly with atoms and molecules, utilizing knowledge of quantum mechanics to probe chemical reactions in real time and in exquisite detail, and even to create new forms of matter. They have also developed ways to investigate the incredibly rapid changes that take place during chemical reactions.

MOVING ATOMS

As if producing stunning, informative images and identifying the chemical elements present on a surface was not enough, scanning probe microscopes can interact directly with individual atoms. By adjusting the tunneling current of an STM and moving to just the correct distance from the surface, for example, the interaction between probe tip and the surface can be "tuned" to just the right level to form a chemical bond between an atom at the probe tip and an atom at the surface (typically, an "adatom": one that has adhered to the surface, or been adsorbed onto it). The bond is strong enough to lift the atom away from the surface and move it, so that it can be deposited at another point.

This remarkable achievement allows for detailed studies of interatomic interactions in the real world—interactions that could previously only be studied mathematically, using quantum theory. But it may also enable nanoscale engineers to build minuscule electronic devices, such as computer memory or quantum dots (see page 141), consisting of just a few atoms, or to fabricate new materials with previously impossible properties. The daunting task of manufacturing such materials atom by atom may be diminished in the future by automating the process. In 2015, researchers at the U.S. National Institute of Standards and Technology (NIST) demonstrated a computer-controlled STM that moved cobalt atoms, strewn randomly on a copper surface, to form a quantum corral (see page 125). They also made quantum dots from carbon monoxide molecules and replicated the NIST logo at the nanoscale—all without human intervention.

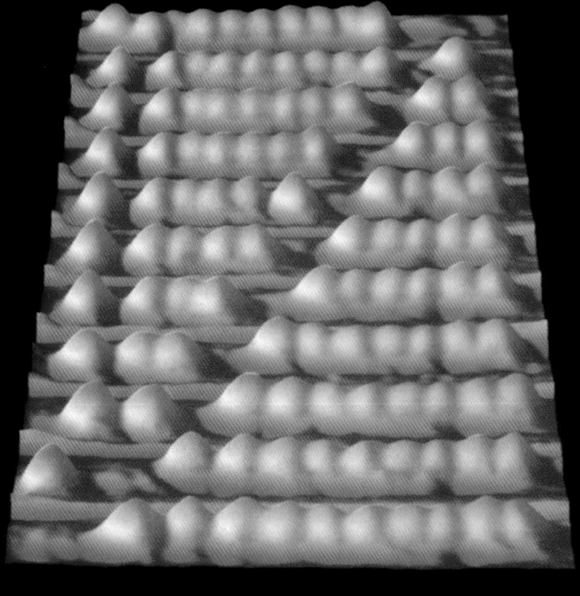

False color STM image of a nanoscale abacus (left). The blue "beads" are buckminsterfullerene molecules (C_{60}, see page 103) and the "frame" is a surface of copper. The STM probe tip was used to pick up and move the beads, one at a time. The false color STM image (facing page) is of cobalt atoms arranged in a circle on a copper surface.

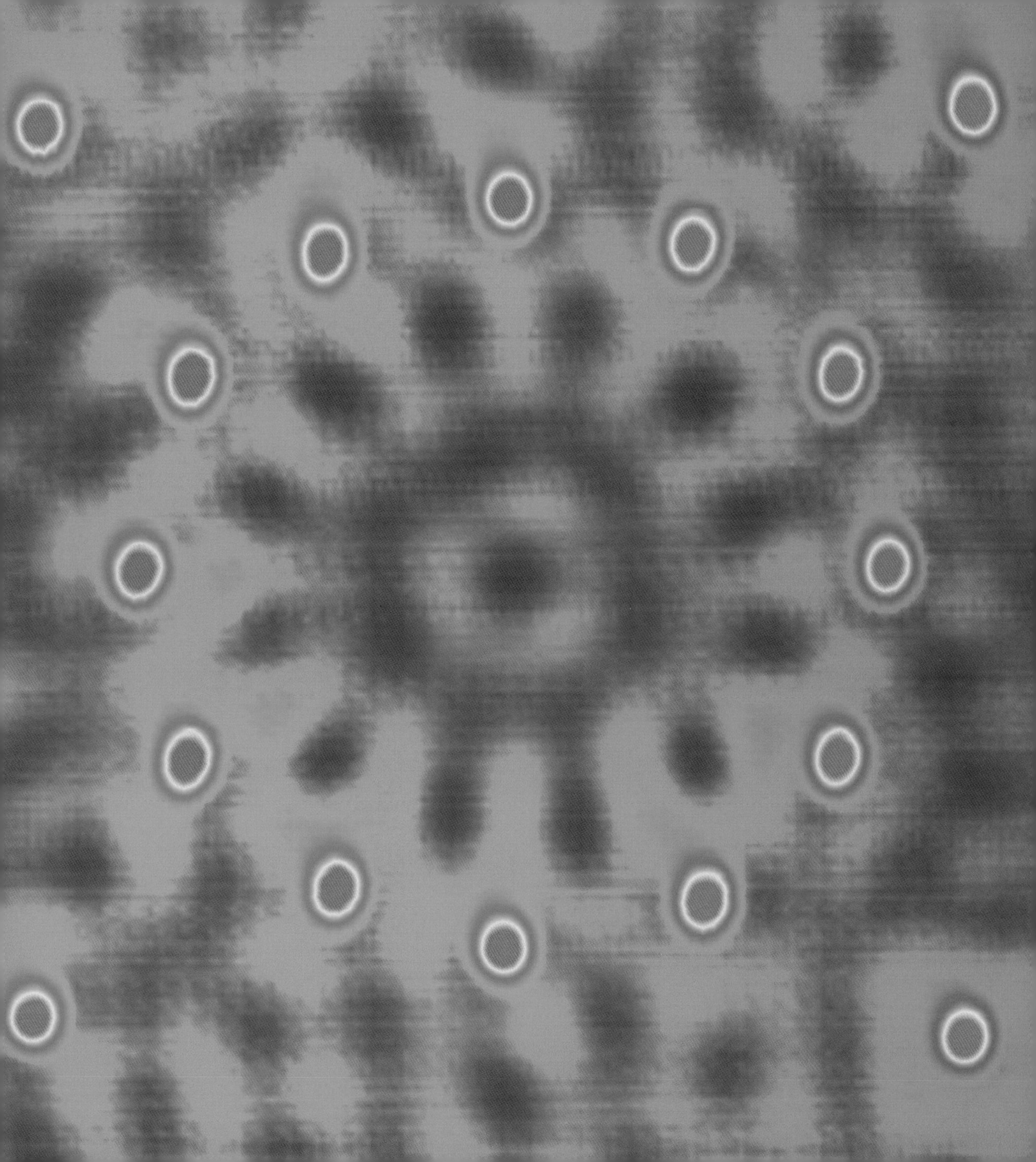

TRIGGERING CHEMICAL REACTIONS

Another remarkable atomic interaction, made possible by a combination of AFM and STM, involved a chemical reaction called a Bergman cyclization (shown on the facing page). The reaction was discovered in 1972 and was initially considered as just a curiosity. But it now shows promise in developing anticancer drugs, because one of the intermediate molecules in the reaction can cleave DNA molecules, making it possible to destroy cancer cells in a targeted way. Researchers at IBM used scanning probe microscopy to study the reaction, and even to make it happen to order. They started by removing some atoms from a single molecule, leaving a stable molecule with three consecutive rings of carbon atoms. They then opened up the central ring, first on one side and then on the other, to make one of two different molecules—by breaking and making bonds between the carbon atoms with a tunneling current, as in STM. The system is a kind of molecular switch that could also one day be used in electronics, the two states of the molecule representing the binary "0" and "1" involved in digital systems.

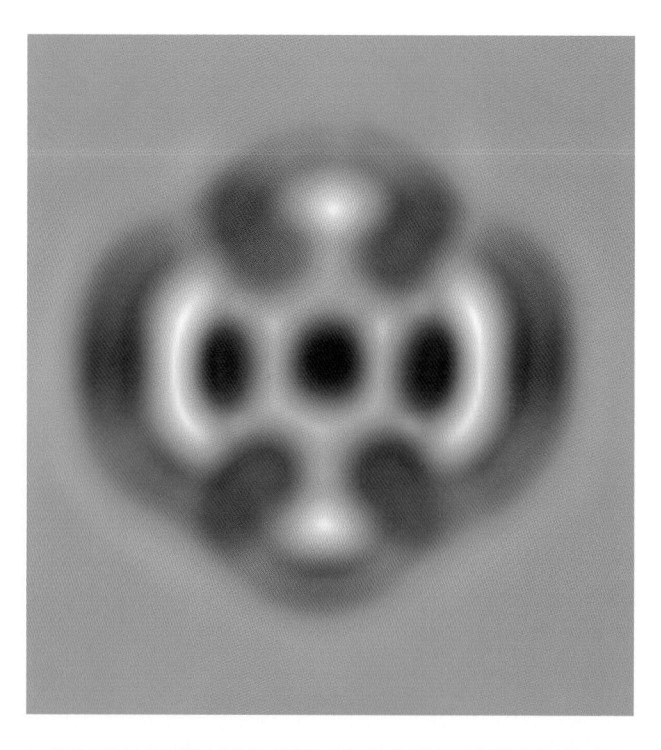

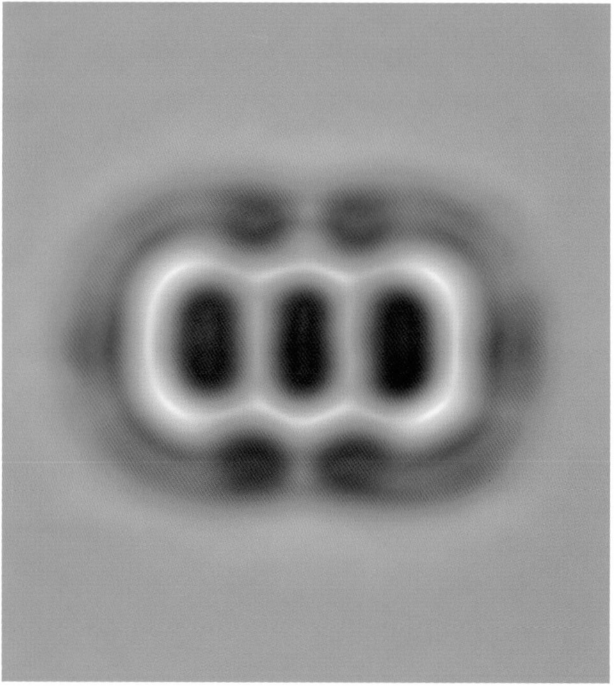

False color AFM image (top) of a single molecule of 9,10-dibromoanthracene on a two-atom-thick layer of sodium chloride. The two bromine atoms are at the top and bottom of the picture. To make this molecule into a molecular switch, IBM researchers first removed the bromine atoms with an STM probe.

False color AFM image (bottom) of the single molecule switch made by IBM researchers after removal of the bromine atoms. The molecule is 9,10-didehydroanthracene. An STM probe tip was then used to break bonds in the molecule, switching between two states (see box on facing page).

BREAKING AND MAKING BONDS

In 2015, IBM researchers made a "switch" with a single molecule of a compound called 9,10-didehydroanthracene. They started with a molecule of 9,10-dibromoanthracene, which has two bromine atoms attached, and began by removing these using the probe tip and tunneling current. The new molecule is only partially stable, and the central ring of carbon atoms can be opened up by breaking the bond between two of the carbon atoms. The researchers placed their molecule on a two-ion-thick layer of sodium chloride at a temperature of −454°F (−270°C). They were able to break the carbon-carbon bond on one side of the central ring or the other, thus switching the molecule between two different states.

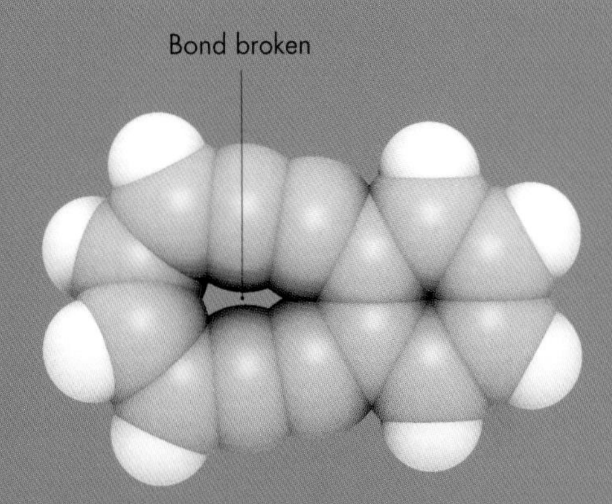

3,4-benzocyclodeca-3,7,9-triene-1,5-diene L

Bromine atom removed

9,10-didehydroanthracene

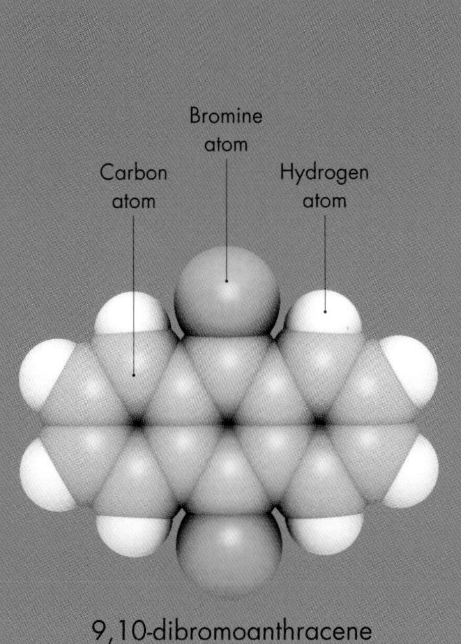

9,10-dibromoanthracene

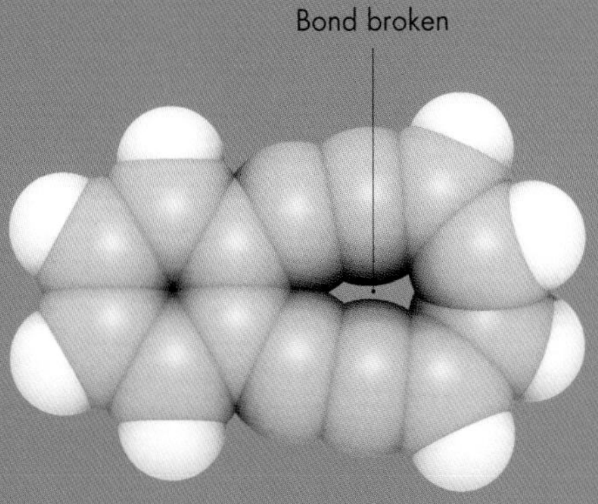

3,4-benzocyclodeca-3,7,9-triene-1,5-diene R

SLOWING DOWN ATOMS

Another way in which atomic physicists can interact directly with atoms is by using lasers and magnetic fields to slow their motion, typically using a device called a magneto-optical trap (see box below). This apparatus can reduce the speeds of a population of atoms dramatically to the extent that they are barely moving and so have almost no kinetic energy. Because the average kinetic energy of a population of particles is related directly to that population's temperature (see page 90), slowing atoms down equates to reducing their temperature. A magneto-optical trap can cool a gas almost to the lowest possible temperature: to -459.67°F (-273.15°C), called "absolute zero." This temperature equates precisely to zero on a specially conceived temperature scale called the Kelvin scale (K). One degree kelvin (1 K) is equal to 1.8°F (1°C), but the Kelvin scale's zero is absolute zero. By interacting with atoms directly, physicists have managed to cool matter down to within one ten-billionth of a degree kelvin of absolute zero. (Note that it is not possible to have zero kinetic energy, because of "quantum fluctuations." These are discussed in chapter seven.)

An actual magneto-optical trap at the Institute of Photonics and Advanced Sensing, University of Adelaide, Australia.

MAGNETO-OPTICAL TRAP

Inside the sealed chamber of a magneto-optical trap, three laser beams meet at right angles to one another. The device's cooling effect depends upon atoms in a gas absorbing photons from the laser beams. It may be hard to imagine, but photons of light have momentum, just as a rolling pool ball does. When an atom absorbs a photon, the photon's momentum is transferred to the atom, just as one pool ball transfers its momentum to another ball if it collides with it. This could either speed up an atom—the opposite of what is intended—or slow it down.

Fortunately, there is a way to guarantee that the absorption will result only in a slowing down. Atoms absorb only certain frequencies of light (see page 48), and the laser frequency is carefully chosen to be a little below one of those frequencies. The atoms will experience the laser light as

The ability to bring temperatures to within fractions of a degree of absolute zero has spawned many insightful research projects into atomic physics and materials science. Most notable among them is the creation of a strange form of matter called a Bose-Einstein condensate (BEC). The potential existence of BECs was first predicted in the 1920s by Albert Einstein and Indian physicist Satyendra Nath Bose. A BEC was created for the first time in 1995 by physicists at the University of Colorado.

In a BEC, all the atoms lose their individual identities and merge into one superparticle. This is because at extremely low temperatures, quantum effects become more prevalent, and atoms behave more like waves than particles; in a BEC, they are all defined by the same wave function (see page 52). At normal temperatures, atoms move around at a wide range of different speeds, and therefore there is a large distribution of different energies. Each atom has a different quantum state (a quantum state is a particle's collection of "observables," such as speed, position, and energy). As the temperature drops to close to absolute zero, the distribution of energies becomes far more restricted—and the quantization of energy comes into play (see page 48). Only certain energies are allowed, and as the atoms lose energy, they "fall" into, and occupy, all the available lowest-energy states—just as electrons fall into the lowest-energy state available in an atom.

having a different frequency, depending upon their direction of movement, thanks to the Doppler effect—most familiar as the heightened pitch of a siren as an emergency vehicle approaches, while you are moving with a relative speed toward the vehicle. An atom moving toward the light will experience the laser's frequency as slightly higher—just enough to make it able to absorb the photon. (For atoms moving away, the same beam reflected off a mirror is approaching in the opposite direction.) The momentum of a single photon is tiny, but, after absorbing many photons, the average speed of the atoms in a gas is reduced from hundreds of meters per second to a few centimeters per second—and the temperature to a tiny fraction of a degree kelvin. A magnetic field permeating the magneto-optical trap chamber confines the cold atoms in an increasingly tighter space.

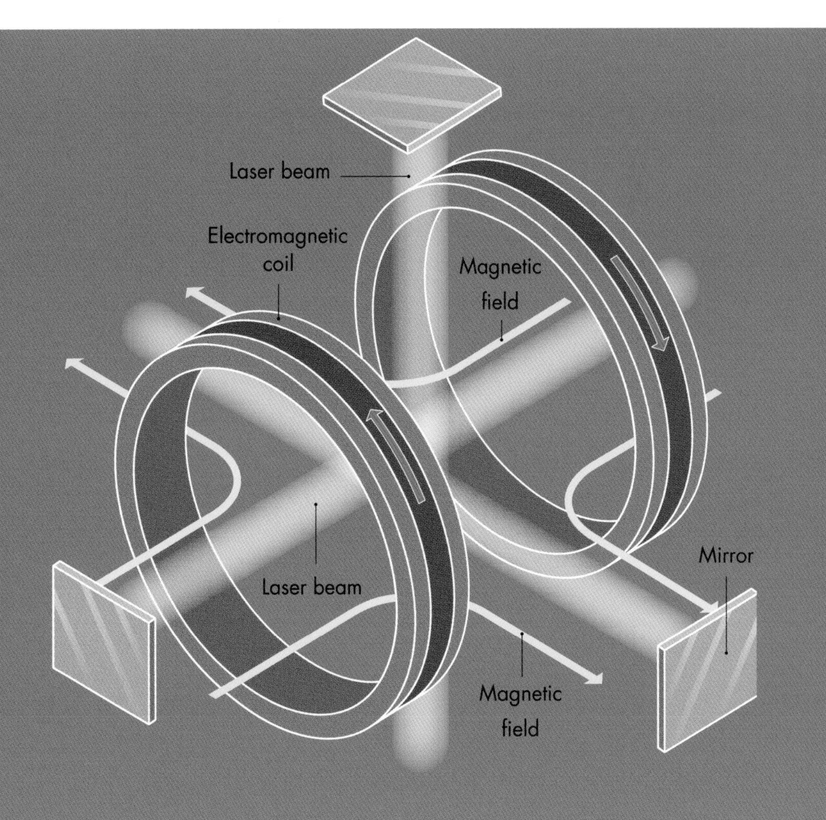

In an atom, no two electrons can have precisely the same quantum state; they fill from the lowest-energy state. This behavior is characteristic of a class of particles called "fermions" (discussed further in chapter seven). Electrons, protons, and neutrons are all fermions, but not all particles are. Some are "bosons," any number of which can occupy exactly the same energy state at the same time. Despite the fact that all atoms contain electrons, protons, and neutrons, some atoms as a whole are bosons. It is those bosonic atoms that can be coaxed into assuming the form of a BEC.

In order to create a BEC, experimenters inject a gas of (bosonic) atoms, normally rubidium or sodium, into the chamber of a magneto-optical trap that has had the air evacuated from it beforehand. The magneto-optical trap cools the gas to a few thousandths of a degree kelvin above absolute zero. The experimenters then irradiate the very cold gas with radio waves, which knocks the fastest-moving atoms away. This evaporative cooling (see page 92) removes only the most energetic atoms, reducing the cloud's average kinetic energy significantly—enough for the temperature to drop to within a few millionths of a degree above absolute zero. Despite the very low temperature, the gas atoms neither condense to form a liquid nor freeze to form a solid, because the gas is rarefied; there are far fewer atoms per cubic centimeter than there are in air, let alone in a solid. Even as the atoms are squashed into a small space at the center of the chamber, they become a BEC before they have a chance to form a liquid or a solid.

FEMTOCHEMISTRY AND ATTOCHEMISTRY

The interactions between lasers and atoms have also allowed atomic physicists and chemists to investigate chemical reactions in unprecedented detail. The breaking and making of bonds that constitute chemical reactions (see page 108) occur very quickly—on timescales of femtoseconds (million-billionths of a second) and attoseconds (billion-billionths of a second). There are about as many attoseconds in one second as there have been seconds in the entire lifetime of the universe, so an attosecond is an extremely brief period of time.

The technology for investigating chemical reactions using lasers was developed in the 1990s by Egyptian-American chemist Ahmed Zewail. He was awarded the Nobel Prize for Chemistry in 1999 for his contributions. The main requirement is that the pulses of laser light used must be as short as possible—just as a camera shutter needs to remain open for as brief a period as possible if the camera is to capture a very fast-moving subject without blurring. This extreme resolution in the temporal (time) dimension is akin to the extreme resolution in the spatial dimensions achieved by scanning probe microscopes and scanning transmission electron microscopes; it presents another way to examine atomic-scale processes in intimate detail. Early femtosecond research, carried out by Zewail and his colleagues, involved first triggering a chemical reaction with a very short laser pulse. They then probed the state of the atoms and molecules taking part in the reaction at very short times afterward, using another short laser pulse. This "pulse-probe" approach is still the basis for investigating reactions on extremely short timescales, but there are now many variations on the theme. Several laboratories have also broken through the femtosecond barrier and are now working with pulses of laser light only attoseconds long.

Before ultrashort laser pulses were available, chemists could only work out how a reaction proceeds from scrutinizing the products—just as crash-scene investigators can only piece together what happened in an automobile accident by examining the wreckage. Femtochemistry and attochemistry provide the opportunity to construct an animation of the reaction

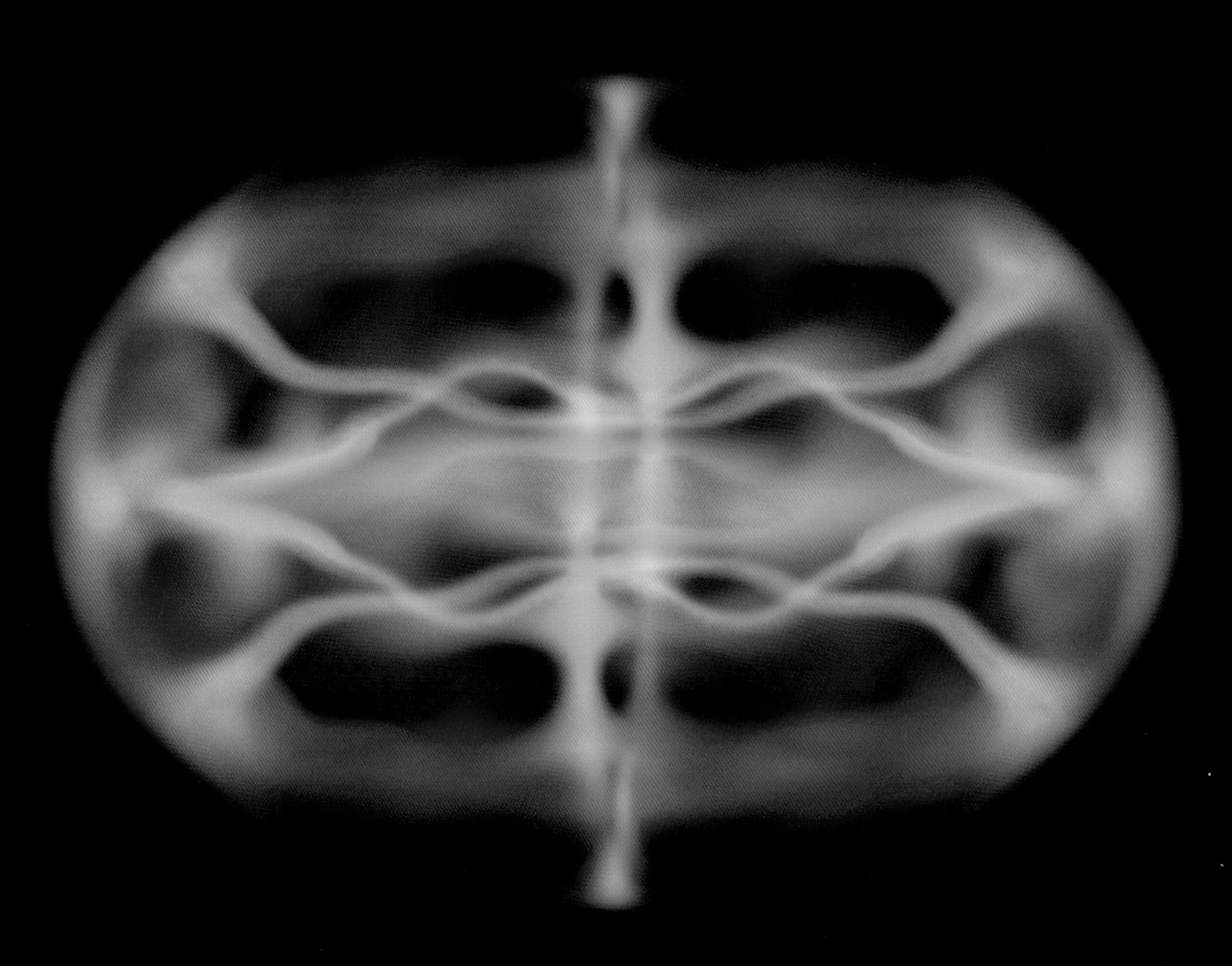

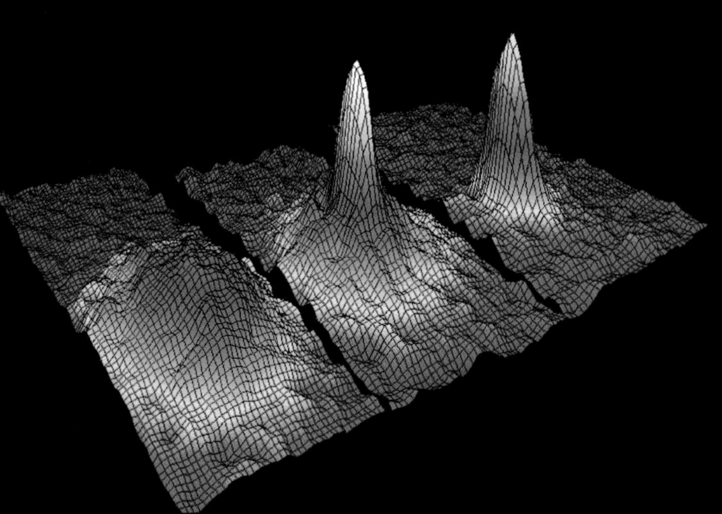

Computer-generated simulation of the wave functions of cold, confined atoms as they begin to merge to form a Bose-Einstein condensate.

Graph showing the formation of the first Bose-Einstein condensate in 1995. The colors represent the density of rubidium atoms inside the magneto-optical trap. Red represents lowest density. White shows where the atoms have coalesced so that they share a single wave function.

in progress—similar to watching a crash over and over in slow motion. Every chemical reaction has an intermediate stage, between reactants and products, in which interatomic bonds have broken but new ones have not yet been made. The intermediate products of a reaction typically only last femtoseconds, or even attoseconds. The main driver of the development of extremely short timescale research is understanding in greater detail the dynamics of how these intermediates form and interact. In pulse-probe femto- and attosecond research, the first laser pulse knocks electrons to higher-energy levels, or provides enough energy to break the interatomic bonds of the reacting molecules. The second pulse interrogates the intermediate products. Typically, the researchers ascertain which bonds have been broken, and which electrons are at what levels, by observing which frequencies of light the molecules absorb—so this is a form of spectroscopy (see page 68). The researchers vary the interval between pulse and probe, enabling them to build up a picture of how the reaction proceeds from one tiny moment to the next.

The range of technologies available to short timescale researchers is expanding, and the basic pulse-probe approach is no longer the only way to interrogate rapid changes in molecules over very short timescales. Many researchers now use "free electron lasers" (see page 145), which can be tuned to produce powerful beams of electromagnetic radiation across a huge range of frequencies, from radio waves to X-rays. For probing reaction intermediates and products, many researchers employ electron diffraction. In this process, incoming radiation knocks electrons out of their bonds of a molecule; the electrons are then diffracted (see page 50) by the atoms present, and produce interference fringes on a screen. Such ultrafast electron diffraction can help researchers discern the positions of atoms in the intermediate molecules of a reaction to within less than an ångstrom.

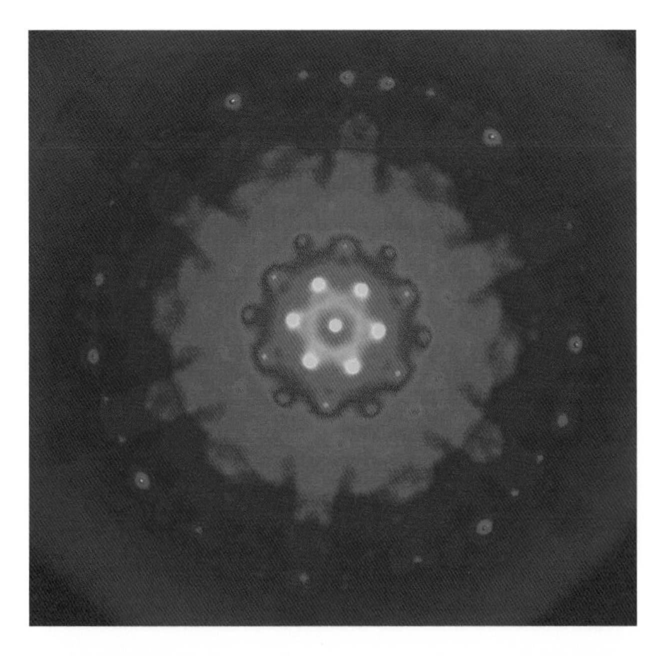

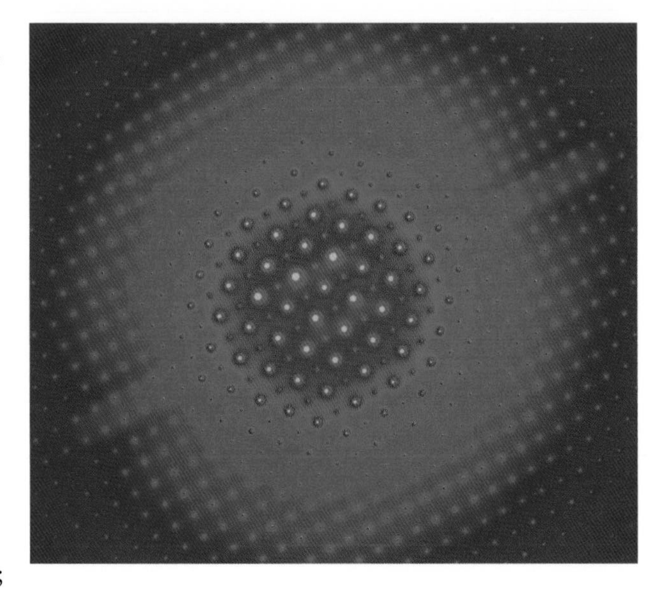

False-colored electron micrographs showing the electron diffraction pattern (see page 115) of pure silicon (top) and molybdenum trioxide (bottom). In ultrafast electron diffraction, incoming electrons disturb the crystal, changing the structure momentarily and revealing information about the bonds between the atoms.

FEMTOSECOND SPECTROSCOPY

Typical laboratory setup for femtosecond research. The time delay between pulse and probe is varied by moving the mirrors back and forth in the optical path of the probe beam, so that the probe takes less or more time to reach the sample.

Laser pulses are extremely short

Path of pulse laser beam

Probe arrives shortly after pulse

Sample

Mirror

Pulse laser

Probe laser

Probe laser beam passes through variable delay

Detector (spectrometer)

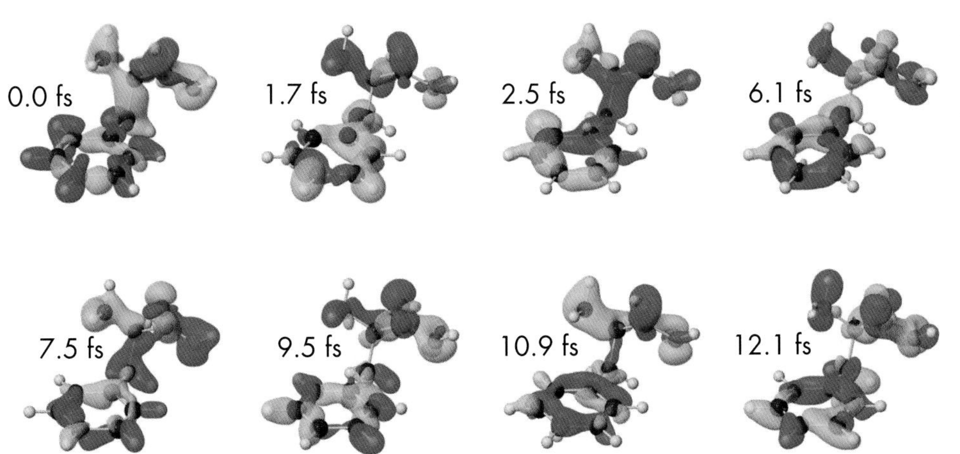

The results of femtosecond spectroscopy reveal extremely rapid changes in molecules. Here, the technique tracks the shifting molecular orbitals in a molecule of phenylalanine after it is hit by a laser pulse. The entire sequence lasts just 12 femtoseconds (fs) or 12 million billionths of a second.

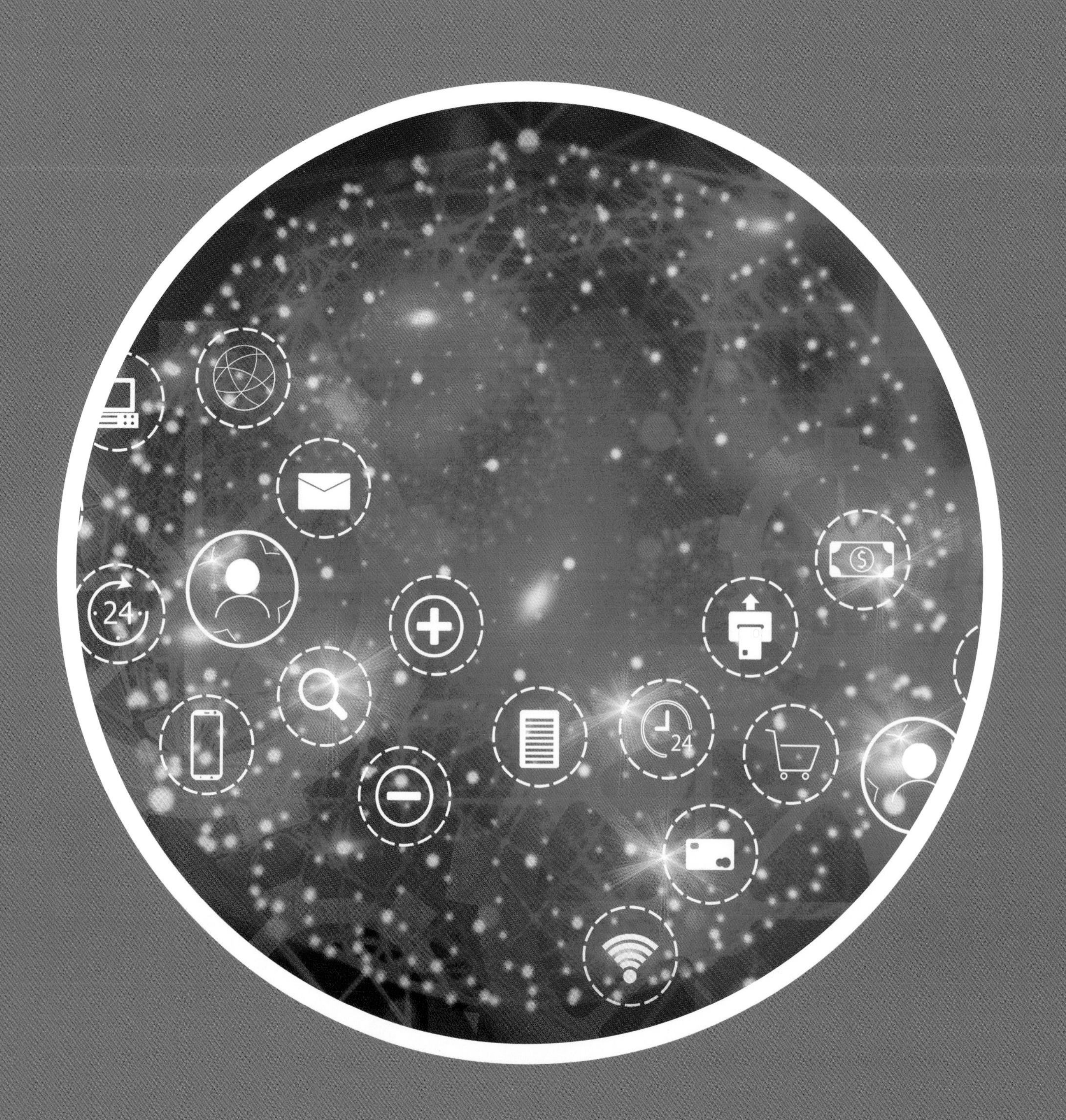
24
24

CHAPTER 6

ATOMIC APPLICATIONS

Many key technologies of the twentieth and twenty-first centuries could not have been invented without an intimate and sophisticated understanding of the workings of atoms. Specifically, without the discovery of radioactivity and the electron, and in particular without quantum mechanics, there would have been no digital revolution and no lasers; no MRI scanners, nuclear medicine, or nuclear power. In this chapter, we will explore and explain some of the technologies whose development has depended upon an increasingly-improving atomic theory.

Many of the things we take for granted in our modern lives could not have been invented without a sophisticated and intimate understanding of how atoms and subatomic particles behave.

SEMICONDUCTOR DEVICES

A semiconductor is an element or compound whose conductivity sits naturally between that of a conductor and an insulator. The conductivity of a semiconductor can be increased when energy is provided in the form of light, heat, or electricity. It can also be manipulated by doping: adding atoms of other elements into the crystal structure. Doped semiconductors are the basis for the electronic components that underpin the digital revolution.

In a good conductor, electrons are able to move around freely within the body of the material. In a metal, this is due to metallic bonding (see page 106). The electrons in the metal atoms' valence shell (the outermost electron shell) occupy the conduction band, a collection of energy levels so similar that they blur together into a continuous range of energies.

Nonmetals also have a conduction band, but the electrons in their valence shell have nowhere near enough energy to occupy it. In other words, for nonmetals, there is a large gap between the "valence band" (the range of energy levels in the valence shell) and the conduction band, which means that the electrons are firmly attached to their respective atoms. Specifically, they are committed to the covalent bonds between the atoms of which the solid is made.

There is a band gap in semiconductor materials, too, but it is much smaller. If you heat a semiconductor, or illuminate it with light, some electrons are given enough energy to be promoted across the band gap and into the conduction band. Even at room temperature, some electrons have enough energy to enter the conduction band, and this explains the fact that a semiconductor has a higher conductivity than an insulator does.

CONDUCTIVITY

In any solid, the energies of the many atoms' electrons are altered by the presence of other atoms, and as a result they merge into "bands" of allowed energies, instead of precise values (see page 107). The electrons involved in bonding have energies that fall in the "valence band." Above a certain energy level, electrons are free of their atoms and can therefore move freely and conduct electricity. These electrons are in the "conduction band." In a metal, the valence and conduction bands overlap, and that is why metals are good conductors of electricity. In an insulator, there is a large gap between the valence band energies and those of the conduction band. A semiconductor has a small band gap, so electrons can easily be promoted by being given just a little extra energy. That is why the conductivity of a semiconductor increases with temperature, for example, or if the material is illuminated by light.

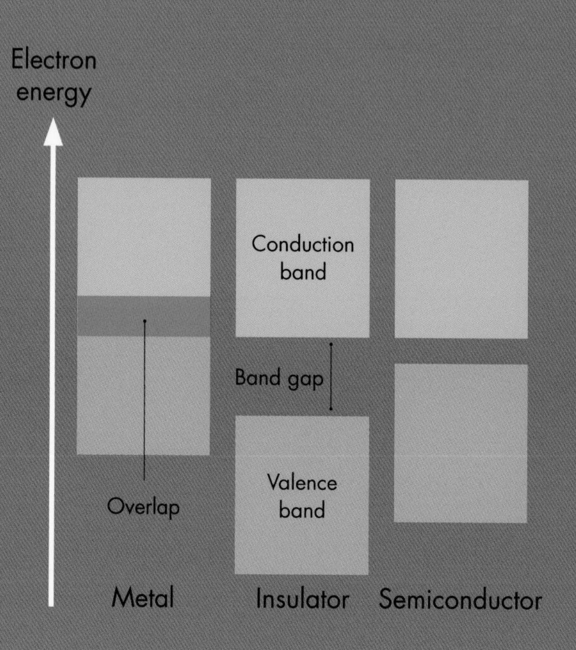

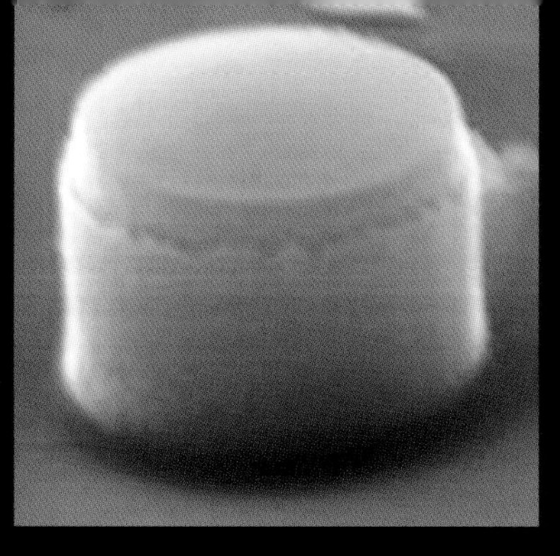

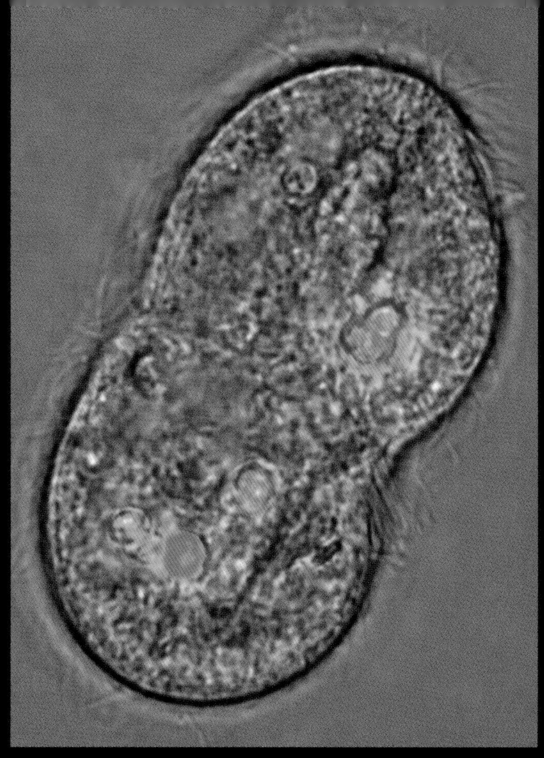

Tiny quantum dots like the two shown here (far left) could play a crucial role in future electronics and communications applications.

In a light microscope image (left) of a breast cancer cell, the cell has absorbed quantum dots that light up when a particular cancer-related protein is active.

An electron is negatively charged, so when it is promoted from the valence band to the conduction band, it leaves behind a point of relative positive charge, referred to as an electron hole, or simply a hole. The electron and hole may recombine, as the electron loses its energy and "falls" back down to the lower-energy level within the valence band. An electron set free in the conduction band can become a mobile carrier of negative electric charge. If an electrical voltage is applied, that election will move, creating an electric current. Incoming electrons can combine with the holes in the crystal.

When electron-hole pairs combine or recombine, the energy they release typically dissipates around the crystal, as vibrations called phonons—but it can instead produce a photon of light. Tiny crystals of semiconductor materials a few nanometers in diameter, called quantum dots, produce particular frequencies of light when excited electrons recombine with holes in the crystal. The color of light they emit depends upon their size. Quantum dots are already being used in some television screens, where they produce bright, pure colors when irradiated with ultraviolet radiation.

Some materials conduct very well, others hardly at all. The electrical conductivity of a substance is measured in units of siemens per meter (Sm^{-1}). At room temperature, a good conductor, such as copper, has a conductivity of about 10 million Sm^{-1}, while a poor conductor, such as sulfur, has a conductivity of a tiny fraction of 1 Sm^{-1}. The conductivity of a semiconductor at room temperature is between these values—about 1,000 Sm^{-1}. However, that figure increases with temperature (or when the semiconductor is illuminated by electromagnetic radiation), because some electrons are given enough energy to jump the band gap and enter the conduction band.

DOPING

The semiconductor that is most often used in electronics is the element silicon. (The element germanium, some compounds of the element gallium, and several other compounds, such as cadmium sulfide, are also commonplace.) Silicon atoms have four electrons in their valence shell, and in a crystal of pure silicon each atom forms four covalent bonds by sharing one electron with each of four other atoms. With two electrons in each bond, each atom is surrounded by a total of eight electrons, a stable, filled shell configuration (see page 80). Doping involves adding elements with different numbers of valence electrons; it increases the conductivity of a semiconductor.

Phosphorus atoms, for example, have five valence electrons; they are pentavalent. The covalent bonds form as usual, but every so often in the crystal—wherever there is a phosphorus atom—there will be one electron leftover, not involved in bonding. This electron has an energy level close to the conduction band, so surplus electrons are easily able to become mobile charge carriers, and move around the crystal. Doping with a pentavalent element results in an n-type semiconductor, so-called because of the negative charge carriers (the electrons).

Doping with a trivalent element—one that has three electrons in its valence shell—has the opposite effect. Add boron atoms into a silicon crystal, for example, and each boron atom forms covalent bonds with the silicon atoms but leaves a hole in the crystal structure, where one of the bonds in a neighboring silicon atom cannot form. This point in the crystal can easily accept electrons from elsewhere, including nearby silicon atoms: effectively, the hole can move around the crystal. Doping with a trivalent element results in a p-type semiconductor, so-called because of the positive-charge carriers (the holes).

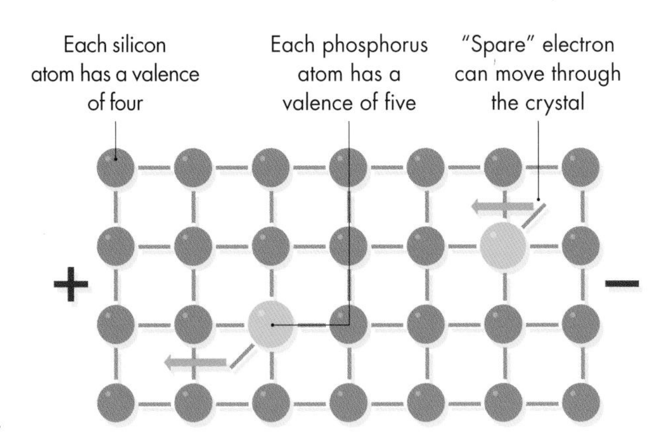

n-type, with phosphorus impurity

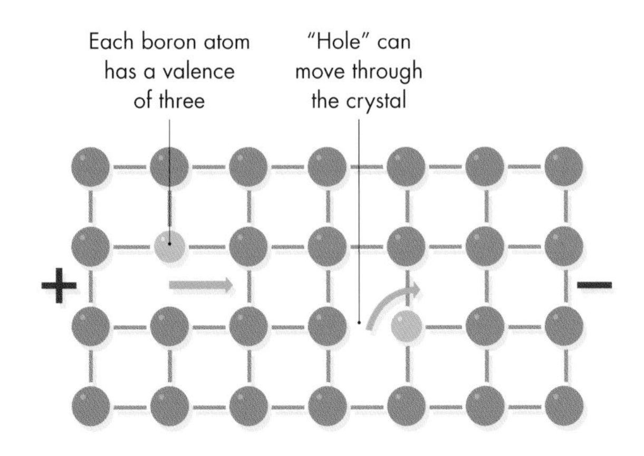

p-type, with boron impurity

Introducing pentavalent atoms into a silicon crystal results in spare electrons, which can move through the crystal and form an electric current. Introduce trivalent atoms instead, and the movable charge carrier is now a deficit of an electron, a positively charged "hole."

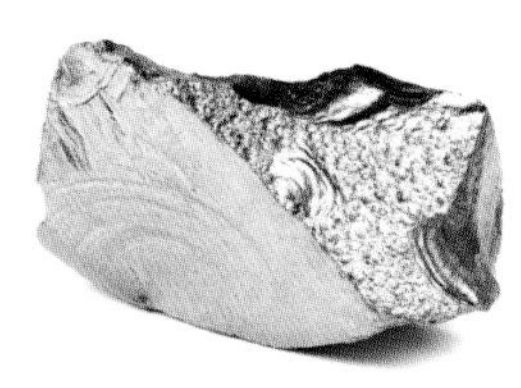

Pure crystalline silicon is a silvery solid at room temperatures. It is the quintessential semiconductor and can easily be doped to produce p- and n-type semiconductors.

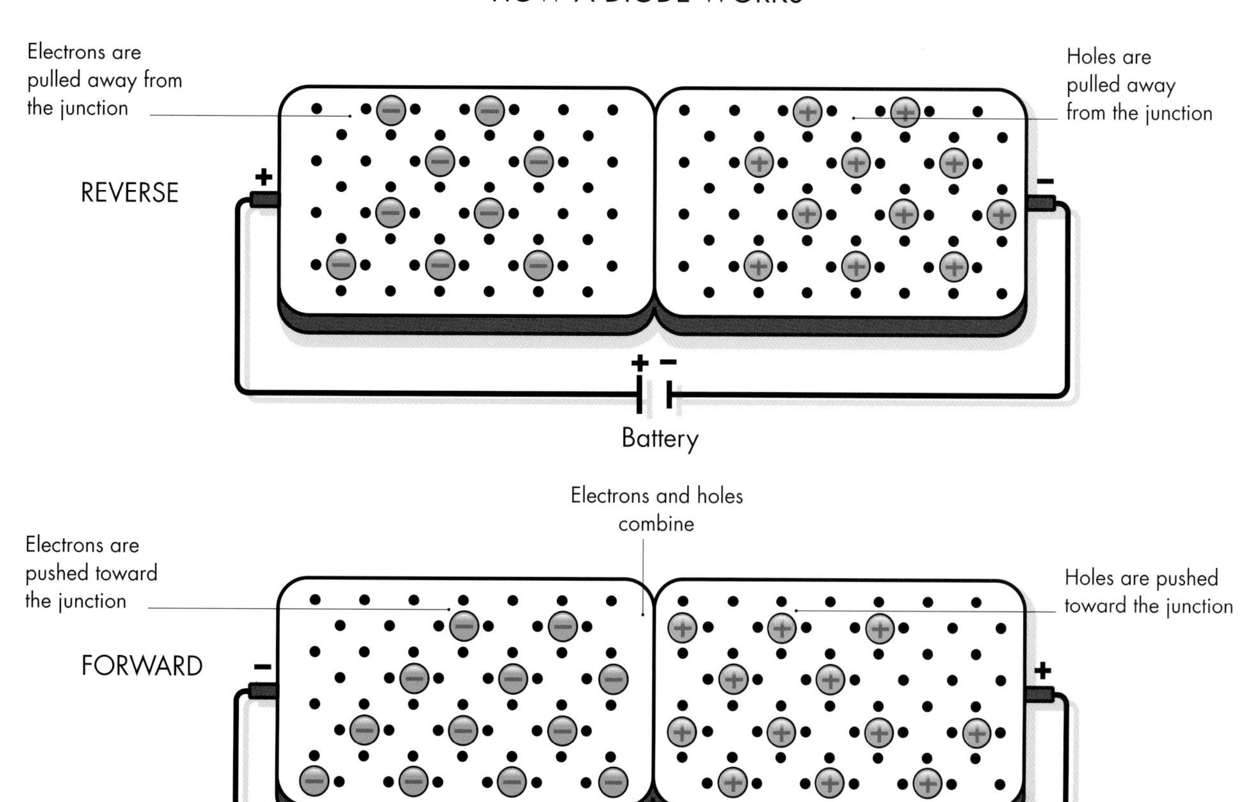

A diode will not conduct electricity when it is connected the wrong way around, in "reverse bias." Electrons and holes are pulled in opposite directions. In "forward bias," the electrons and holes come together and recombine, and the battery can supply more of each charge carrier, so a current can flow.

When placed together, to form a "p-n junction," these two kinds of doped semiconductor become central players in the digital revolution.

DIODES

The p-n junction is the basis of the most fundamental semiconductor-based electronic component: the diode. Connect the negative terminal of a battery (shown above) to the n-type end of the diode, and the positive terminal to the p-type end, and a current will flow through the diode. Electrons are pushed away from the negative terminal, and they cross the junction, where they combine with holes. At the same time, more holes are created in the p-type semiconductor, as electrons are pulled away from that end of the diode by the battery's positive terminal. Connect the battery the other way around, however, and no current flows. This is the essential property of a diode: electricity can only flow through it in one direction.

With careful choice of materials and doping, light of a specific frequency is produced when the electrons combine with holes at the junction. The diode then becomes a light-emitting diode, or LED. These components are found in many electronic devices, often as displays, and also in low-energy lamps. LEDs made with carbon-based compounds that behave like semiconductors are called organic LEDs (OLEDs), and these are common in smartphone displays.

Semiconductor junctions are also at the heart of a laser diode. These cheap, low-power lasers are found in many electronic devices, including bar code readers, DVD players, and laser printers. Like any laser, a laser diode produces a beam of light whose waves are all in phase (meaning they are in step, with the waves' peaks and troughs coinciding). Lasers are another technology whose development has depended upon an understanding of quantum physics (see box on the facing page).

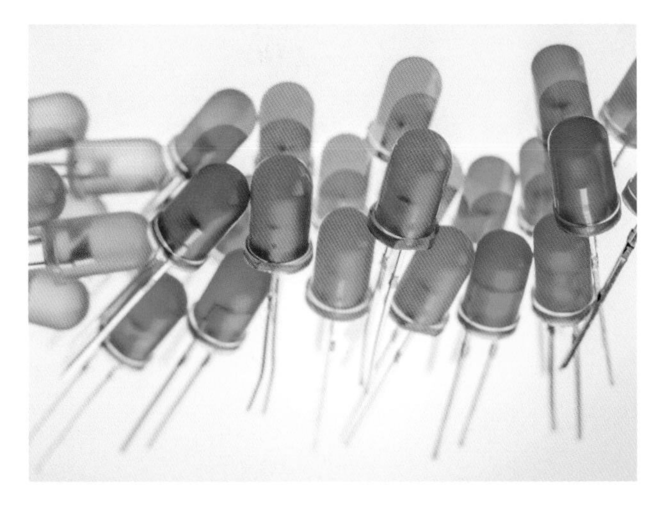

LEDs (above) are cheap and versatile sources of light that are also very energy efficient and that can be switched on and off very quickly. Many modern televisions have an LED panel that provides a white backlight for the screen. An OLED (organic LED) TV screen is different: it has tiny red, green, and blue LEDs that produce the image (below). These low-power screens are flat and can be flexible.

LASERS

The word "laser" comes from the phrase "light amplification by the stimulated emission of radiation." Just as in an ordinary LED, or in a fluorescent lamp, each photon of light produced by a laser diode is created from the energy lost when excited electrons fall down to a lower-energy state. In an LED or fluorescent lamp, the emission of the photons occurs randomly, but in a laser diode, photons are created in concert; the undulations of the light waves are all in step.

The key to the laser diode is a region of undoped semiconductor that sits between the p-type and the n-type regions. As electricity flows through the diode, huge numbers of electrons become excited in this central region. This situation is called a "population inversion" (because normally, the electrons would randomly assume their lower-energy state). The excited electrons remain in their higher-energy state (1) until a photon passes by and triggers their fall to the lower-energy level (2). Each new photon whose creation is triggered by the passing photon is perfectly in step with the incoming one (3). This "stimulated emission" therefore has the effect of amplifying the light; there are now two identical photons where before there was one.

Other kinds of laser work in the same way—with a population inversion, then amplification by stimulated emission—but using something other than a semiconductor as the "lasing medium," typically a solid crystal or a gas. The most versatile type of laser is the free electron laser, in which the light is produced by electrons zigzagging back and forth inside a chamber. This laser is versatile because it can be tuned to produce a huge range of frequencies of light.

Structure of a semiconductor laser (below) and the general working principle of lasers (right).

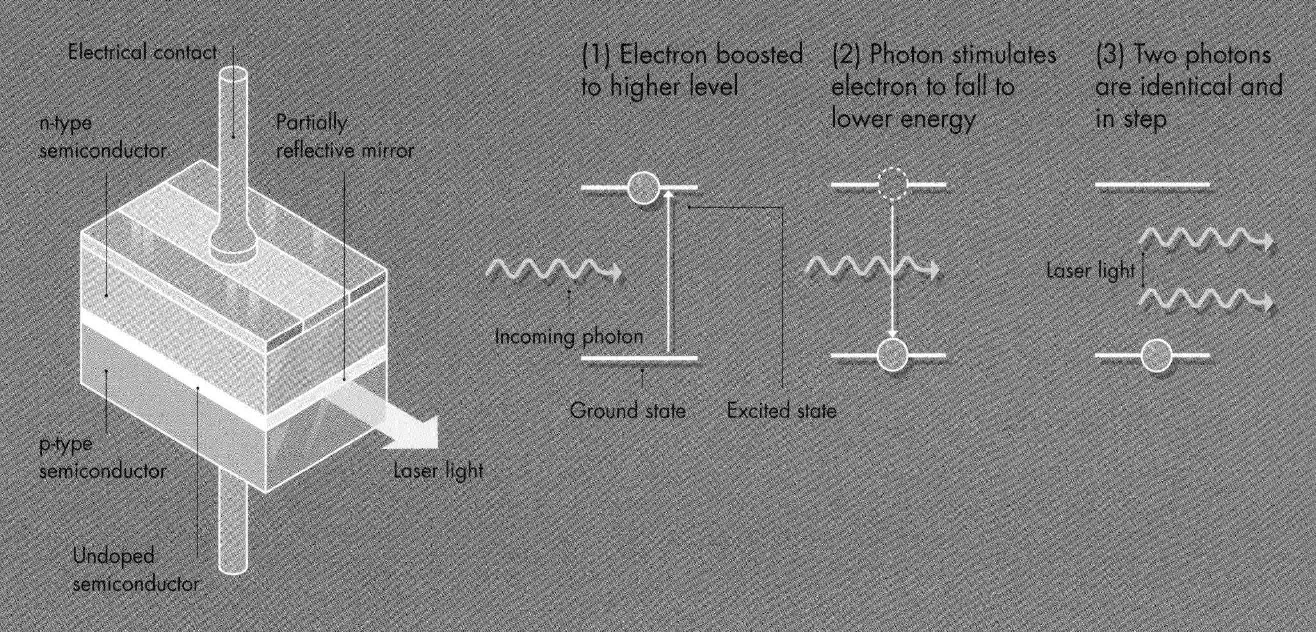

TRANSISTORS

The quintessential semiconductor component is the transistor—and a single microprocessor inside a computer contains a few billion of them. A transistor has two main abilities. First, it can produce large, varying electric currents that are a copy of much smaller input currents. In other words, a transistor can amplify a signal. Second, it can be in two distinct states—"on" (conducting electricity) or "off" (not conducting)—without switching the polarity of the power source, as you would have to do with a diode. These "on" and "off" states can represent the two digits of the binary number system, "0" and "1." This is why the transistor is at the heart of the digital revolution.

The most common type of transistor, especially in the microprocessors in computers and other digital devices, is the field effect transistor (FET). The main current through the transistor flows from one contact, called the source, to another, called the drain. That current can be controlled—typically turned on or off—by an electric field applied at a contact called the gate. Imagine the current through the transistor as the flow of water through a hose, and the electric field at the gate as someone pinching or relaxing a finger and thumb on the hose to shut off or allow the flow of water. The most common type of FET is the metal oxide semiconductor field effect transistor (MOSFET). The name derives from the fact that the gate is typically made of metal insulated from the main body of the transistor by a layer of another material, normally silicon dioxide.

A MOSFET consists of a piece of n-type or p-type semiconductor, with the source and drain being made of the opposite type of semiconductor (p- or n-type). This is the equivalent of having two p-n junctions side by side, each the mirror image of the other. It is like connecting two diodes back to back, so no current can flow through the transistor. Applying a voltage to the gate creates an electric field inside the body of the transistor, which draws electrons or holes toward the gate (depending upon the voltage). The charge carriers (the electrons or holes) now present a channel through which current can flow, from source to drain. This switching process can take place millions of times every second, inside each of the billions of transistors inside a microprocessor. This is how computers manipulate the binary "0"s and "1"s that represent instructions and digitized data, such as alphanumeric characters, sounds, or images.

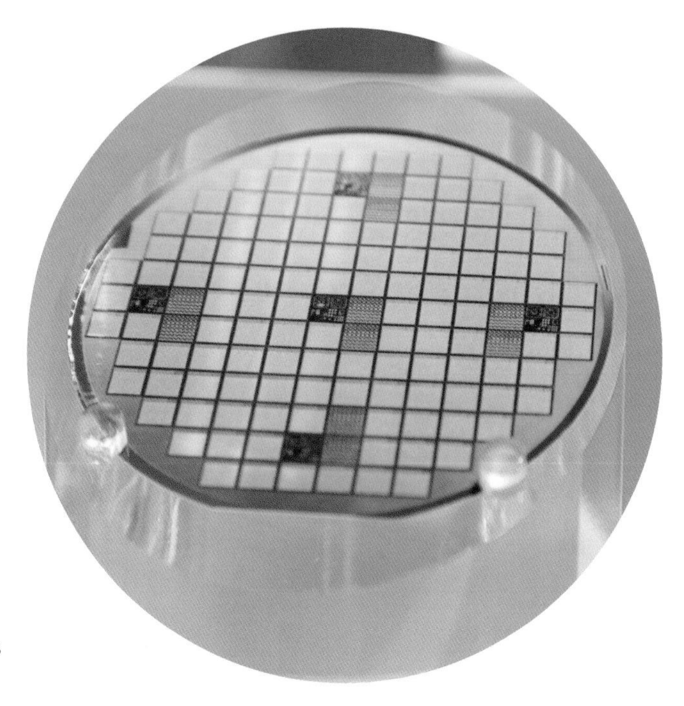

Several dozen square microprocessors are made at the same time on a single circular wafer of silicon. Each microprocessor contains billions of tiny MOSFET transistors.

HOW A MOSFET WORKS

A metal oxide semiconductor field effect transistor is composed of a "bulk" of doped semiconductor with two other regions of doped semiconductor inside it. In these illustrations, the bulk is p-type semiconductor (red), and the other two regions are n-type (green). There are three electrical connections: the source, connected to the negative terminal of a battery; the drain, connected to the positive terminal; and the gate, which can be negative, positive, or have no voltage at all.

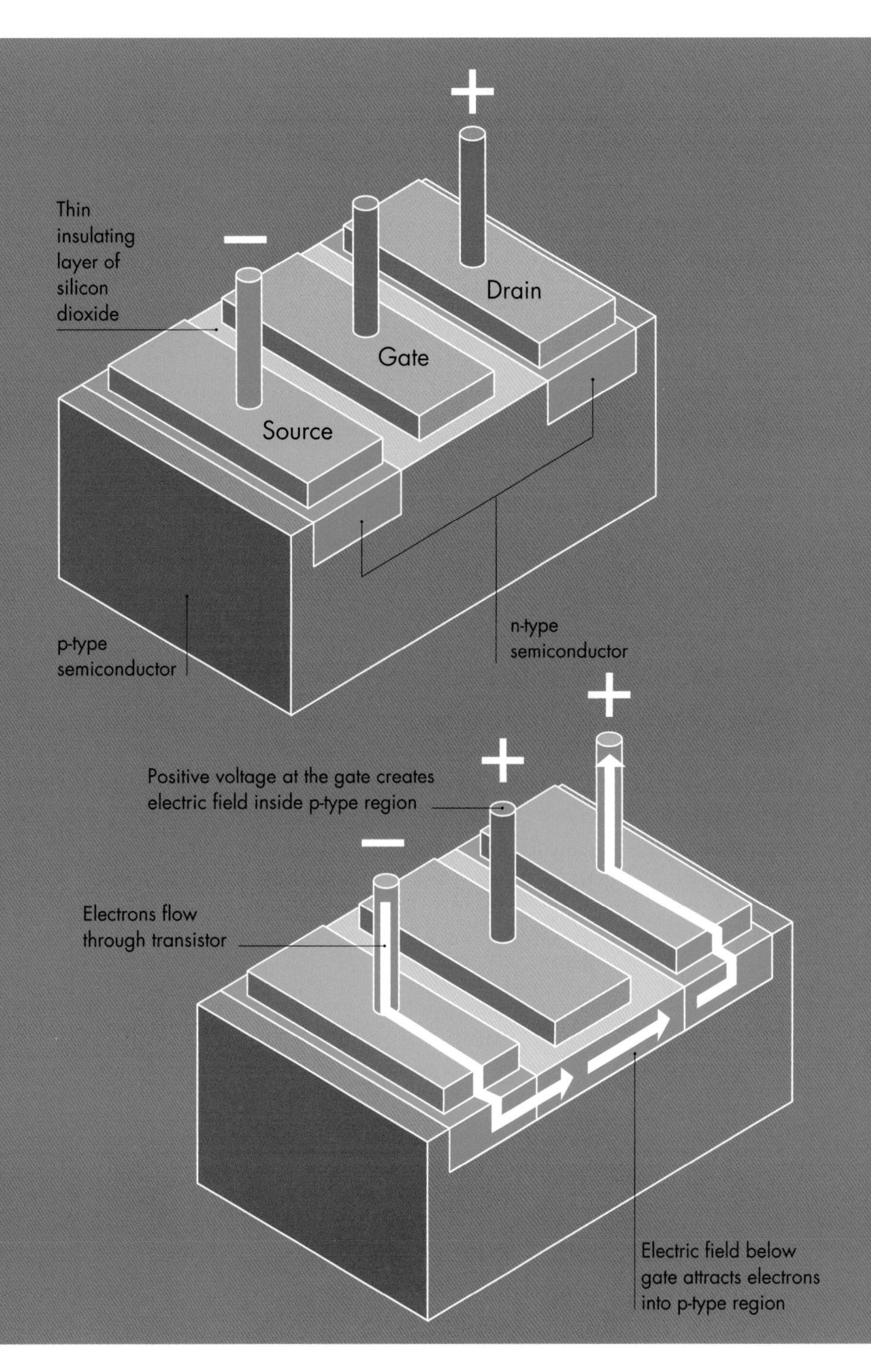

In the first of the two examples, the transistor is "off." No current flows through it, because no electrons are available in the bulk between the two n-type regions. In the second example, there is a positive voltage at the gate. This creates an electric field inside the p-type bulk, and that field draws electrons up toward the gate. (Electrons cannot flow into the gate itself, because there is an insulating layer of silicon dioxide between it and the p-type bulk.) The electrons now present near the gate are the right kind of charge carrier for the two adjacent n-type regions, and a continuous stream of electrons can flow from source to drain. The transistor is "on."

MAGNETISM

People have been experimenting with magnetism for at least two thousand years. However, it is only in the past century or so that scientists have come to understand how magnetism works, and to exploit its potential more fully. The magnetism that we are familiar with depends largely upon a property of electrons called "spin." But some more recent technologies depend upon the "spin" of the nucleus, not electrons—most notably, the MRI scanner.

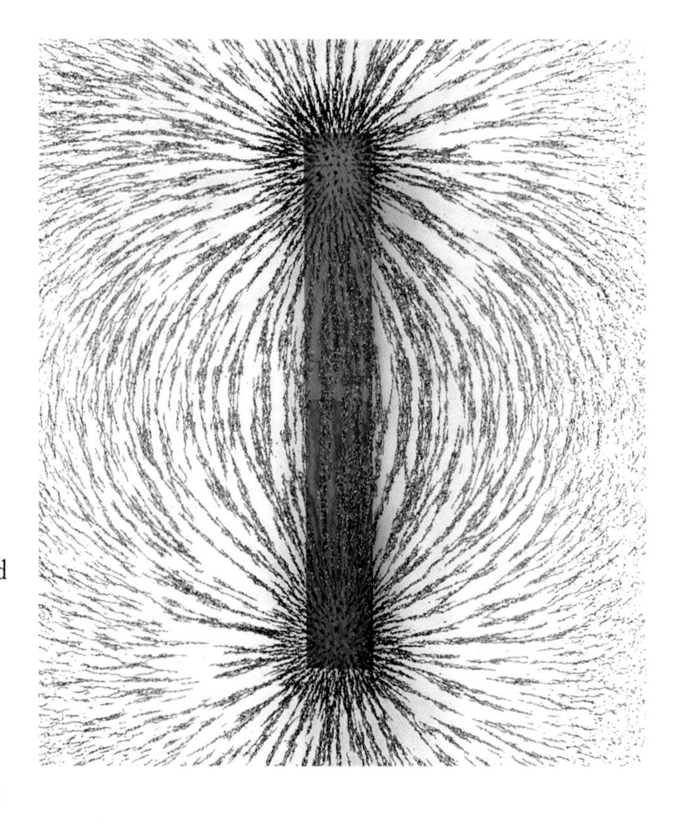

Iron filings become magnetized in the magnetic field of a permanent magnet and are attracted to each other, highlighting the field itself. Inside a magnetic material, such as iron filings (shown above), tiny magnetic regions called domains are normally randomly oriented (shown below left). In a magnetic field, they align (below right).

WHAT IS MAGNETISM?

The force that makes a magnet cling to a refrigerator door and the force that turns an electric motor are both manifestations of magnetism. Magnetic forces are carried by magnetic fields. The field of the refrigerator magnet is the combination of all the weak magnetic fields of tiny crystals of which the magnet is made, or domains (see below). Inside the motor, there are two interacting magnetic fields, at least one of which is generated by electric currents in a coil of wire. The electrons coursing through the wires in the motor are less relevant to this book than the refrigerator magnet, whose magnetism is the result of electrons inside atoms. Nevertheless, there is a deep connection between the two examples. In both cases, the magnetic fields are produced by particles that carry electric charge—something that is true of every magnetic field.

Unmagnetized

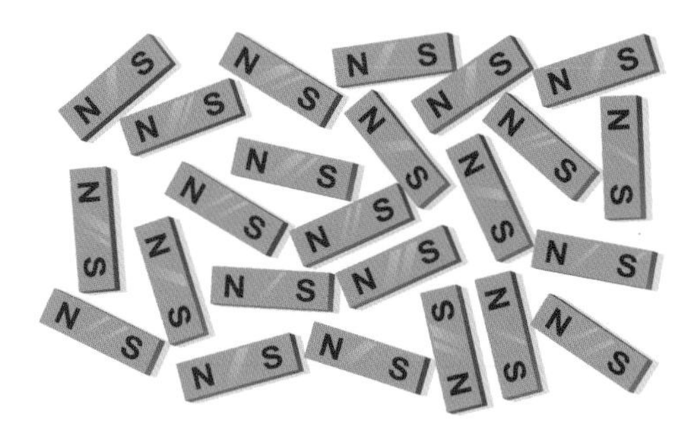

Magnetized

N S N S N S N S
N S N S N S N S
N S N S N S N S
N S N S N S N S
N S N S N S N S
N S N S N S N S

A single atom can have its own magnetic field—the combination of all the magnetic fields of its constituent parts. (In some atoms, the contributions of the nucleus and the electrons all but cancel out, but most really are tiny magnets.) The electrons contribute by far the biggest part of the magnetic field around an atom. Their orbital motion around the atom's nucleus is the equivalent of a looping electric current, like the current in the electromagnet coils in the motor. Note that, according to quantum mechanics, electrons exist as clouds of probability (see page 34), and are not really revolving around the nucleus as planets orbit the Sun—but they have the same effect as if they were orbiting.

The electrons also have a quality called "spin," which also gives them their own magnetic field. The term was coined as the result of experiments in the 1920s, which showed that electrons have magnetic effects beyond their orbital motion. Physicists suggested that the electrons were small spheres of electric charge that spin on their axis, just as planet Earth spins. It turns out that such an explanation does not tally with reality. Instead, "spin" is a strange, intrinsic property of all subatomic particles that is not easily understood outside of the mathematics of quantum field theory (a subject discussed further in chapter seven).

Iron is the most common magnetic material. A piece of iron (or steel, which is mostly iron) readily clings to a magnet. The fields of the domains (the tiny crystals within the iron, within each of which the iron atoms' magnetic fields are aligned) are randomly oriented, so the iron has no magnetic field overall. But the presence of a magnetic field within the iron aligns the magnetic fields of the iron's domains in the same direction, so that the iron becomes magnetized. Then the permanent magnet and the iron cling together, their fields interacting.

SPIN

The quantum mechanical property called spin, or, more properly, spin angular momentum, does not refer to a rotation, such as a spinning top. It is an intrinsic property of all particles. Like energy, spin is quantized; it can only have certain values. The spin of an electron may be either +1/2 or −1/2 (in carefully chosen units appropriate for quantum mechanics), or "spin up" and "spin down." Electrons sharing the same orbital are in exactly the same quantum state apart from their spin—that is why two electrons are allowed in each orbital, differentiated by only their opposing spins. If an orbital is filled, with its complement of two electrons, the magnetic fields of the two electrons cancel out. But a lone electron in an orbital—an unpaired electron—does have a magnetic field. Other particles also have spin. Protons and neutrons, for example, also have half integer spin (−1/2 or +1/2). Their spin is the result of the spin of the quarks and gluons of which they are made (see page 39). The photon has a spin of 1, but it has no electric charge, so it does not have a magnetic field.

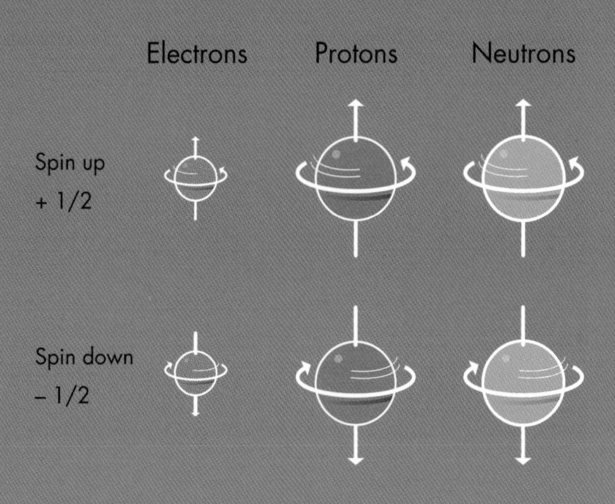

Iron has magnetic domains because its atoms are magnetic; the atoms have unpaired electrons. In most atoms, any unpaired electrons pair up via ionic or covalent bonds (see page 98), but the unpaired electrons in iron (and many other metals) are in energy levels below the valence shell, so do not take part in bonding. These electrons remain unpaired, and their spins align, giving the atom a (relatively) strong magnetic field.

A better understanding of the causes of magnetism, and of the relevance of electron configuration to the structure of the periodic table, led to the development of stronger magnets during the twentieth century. In particular, from the 1950s, magnet manufacturers began to use rare earth metals from the f-block of the periodic table (see page 78). Neodymium is the most commonly used magnetic rare earth metal. It is used, in an alloy with iron and boron, in hard-disk drive motors as well as the generators in wind turbines and the motors of electric cars. Neodymium atoms have seven unpaired electrons—and neodymium magnets are the strongest permanent magnets available. They can therefore be much smaller than other magnetizable materials for their size.

The strong permanent magnets inside wind turbines are made of a neodymium-iron-boron alloy. When the turbine blades spin, the magnets turn around fixed coils of wire, inducing a current in the wires.

NUCLEAR MAGNETISM

Atomic nuclei can act as tiny magnets, too, because the protons and neutrons of which they are made have electric charge and spin. Within a nucleus, protons pair up with other protons of the same energy, just as electrons do in their orbitals, and neutrons pair with other neutrons. Each member of a pair has the same quantum state as the other, apart from spin: one "up," one "down." The magnetic fields of two paired protons or two paired neutrons cancel out, as is the case for paired electrons. So any nucleus with an even number of protons and an even number of neutrons will have no magnetic field overall. But for many nuclides that is not the case; they do have a magnetic field. The magnetic field of a nucleus is much weaker than the magnetic field of even a single electron, but in certain circumstances it can be used in a productive way.

Put a material in a strong magnetic field and the spins of most of the nuclei will line up with the field. In addition, each nucleus will "precess," which means its spin axis describes a circle as a gyroscope or a slowing spinning top does. The frequency of the precession depends upon the total number of protons and neutrons in the nucleus and the strength of the external magnetic field. If you now irradiate the material with a pulse of electromagnetic radiation with the same frequency as the precession, many of the nuclei will change the direction of their spin axis by either 90 or 180 degrees. Because this only happens when the frequency of the radiation is matched to the frequency of precession, it is an example of resonance. One everyday example of

NUCLEAR SPIN

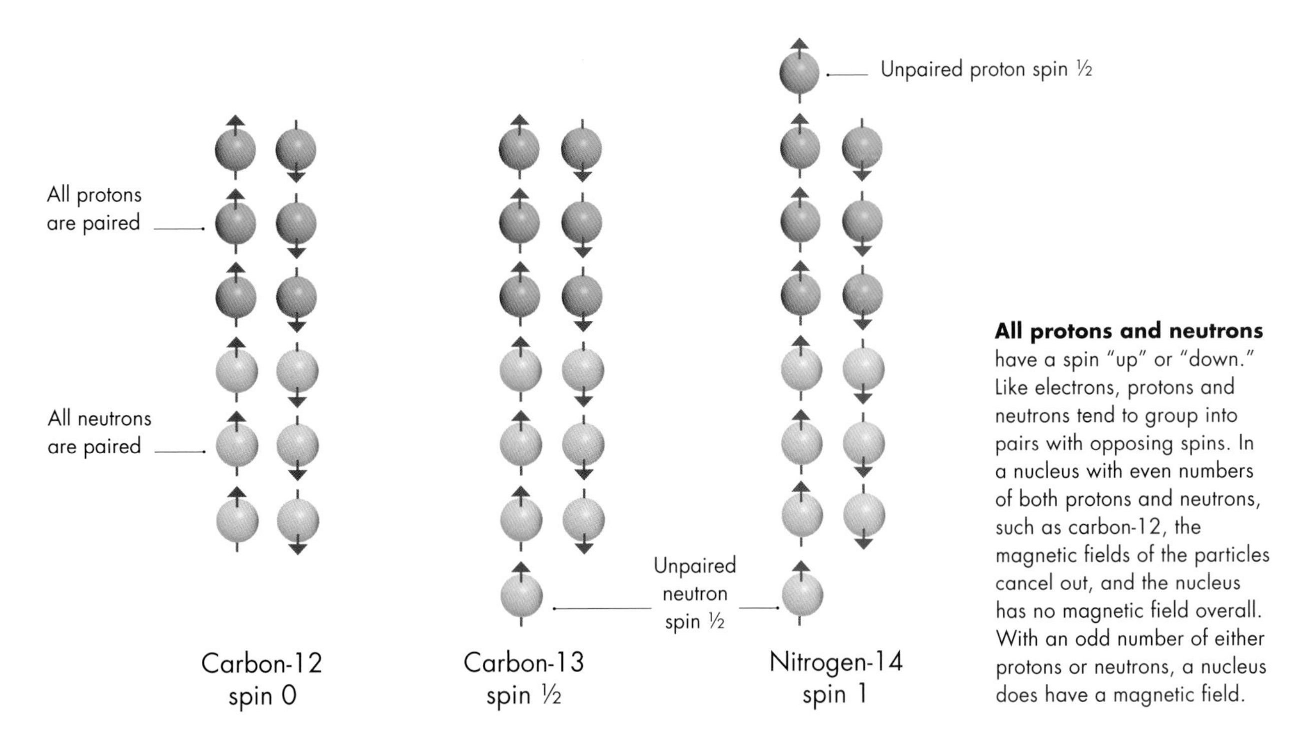

All protons and neutrons have a spin "up" or "down." Like electrons, protons and neutrons tend to group into pairs with opposing spins. In a nucleus with even numbers of both protons and neutrons, such as carbon-12, the magnetic fields of the particles cancel out, and the nucleus has no magnetic field overall. With an odd number of either protons or neutrons, a nucleus does have a magnetic field.

resonance occurs when you push someone on a swing, pushing at just the right moment each time to increase the height the swing reaches. Another example of resonance is vibrating a wine glass to the point of destruction by singing a loud note. Flick a wine glass and it will vibrate at a certain frequency, which is determined by its size and the thickness of the glass. The wine glass will vibrate to destruction only if you sing a note at the "resonant frequency." The process of flipping the nuclear spins similarly happens only at the nuclei's resonant frequency; it is referred to as nuclear magnetic resonance (NMR). The radio waves hitting the nucleus not only cause some nuclei to flip, but they also make the nuclei precess together (in phase), instead of randomly.

Once the radio wave pulse has ceased, the nuclei begin to relax back to their starting position. They do not all flip back at the same time, but one by one, randomly. As they do so, the nuclei give out a radio signal. The time it takes for all the nuclei to relax back depends upon the type of material the nuclei inhabit, and this allows for the differentiation between types of material. Another kind of "relaxation" occurs after the radio pulse ceases. Then the precessing nuclei, synchronized by the radio waves, begin to lose their synchronization, and to precess out of phase again—partly because of interactions with other atoms nearby. When they are in phase, the precessing nuclei emit a strong radio signal, which diminishes as the synchronization is lost. The time it takes for this relaxation to occur also depends upon nearby atoms and is therefore characteristic of the material being tested. The most important and widespread use of NMR is in magnetic resonance imaging (MRI). This produces detailed three-dimensional images of the internal anatomy of the body, with different tissue types clearly visible (see the box on the following page).

MAGNETIC RESONANCE IMAGING

Radiologists in hospitals use nuclear magnetic resonance in MRI scanners. A strong magnetic field aligns magnetic nuclei and causes them to precess, and a series of radio wave pulses flips them. The resonant frequency is chosen to be that of the nuclei of hydrogen atoms, which are extremely common in living tissues. The time it takes nuclei to relax back from their flipped state, and the time it takes them to lose their precessional synchronization, are characteristic of the tissue type. So are the variations in density of hydrogen in different tissues (in particular, watery tissues have a high concentration of hydrogen, resulting in a stronger signal overall).

There is also a second set of magnetic fields, oriented at right angles to each other, and their strength varies along the length, breadth, and depth of the patient. This variation creates a slightly different resonant frequency at every point inside the body, which gives a slightly different signal from each point. A computer analyzes the signals and pieces together a three-dimensional image that shows what kind of tissue is at what point in space.

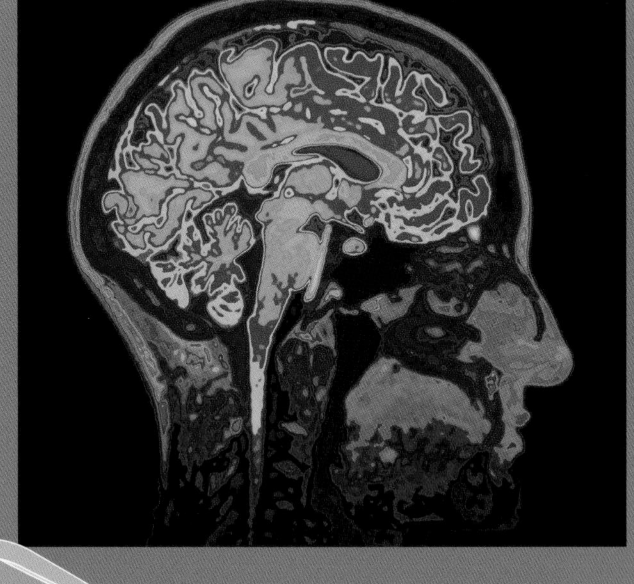

A false-color MRI "slice" through a human head shows the different types of tissue. Bone is blue; fatty neuron-rich brain tissue is orange and red; other tissues are purple.

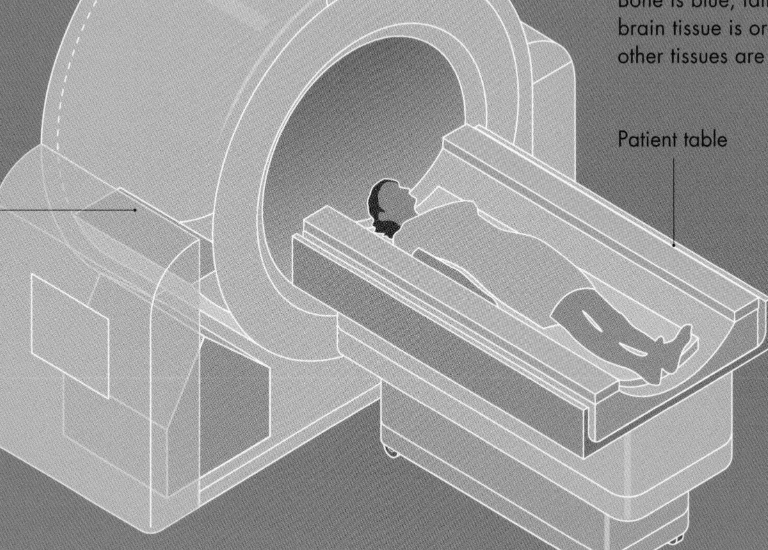

Diagram of an MRI scanner
The main magnet coils are bathed in liquid helium, which allows current to flow freely in them. They are superconducting coils and produce a very strong magnetic field.

(1) No magnetic field

Hydrogen nucleus (proton)

Spins oriented randomly

(2) Strong magnetic field inside MRI machine

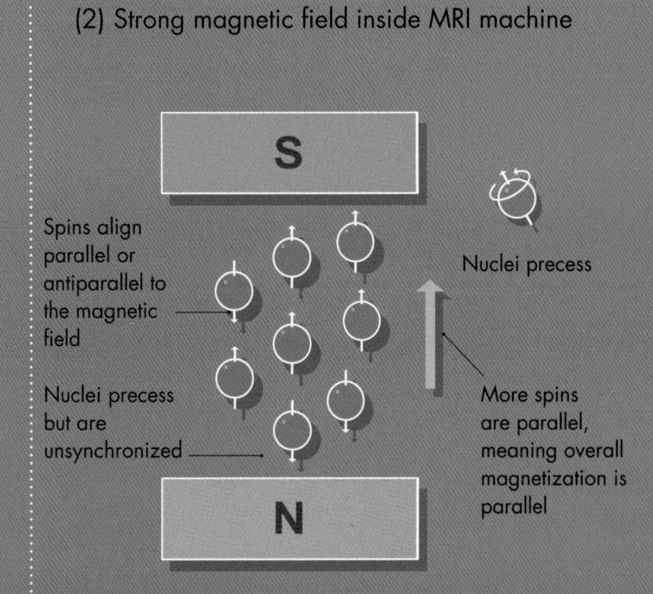

(3) Pulse of radio waves at resonant frequency

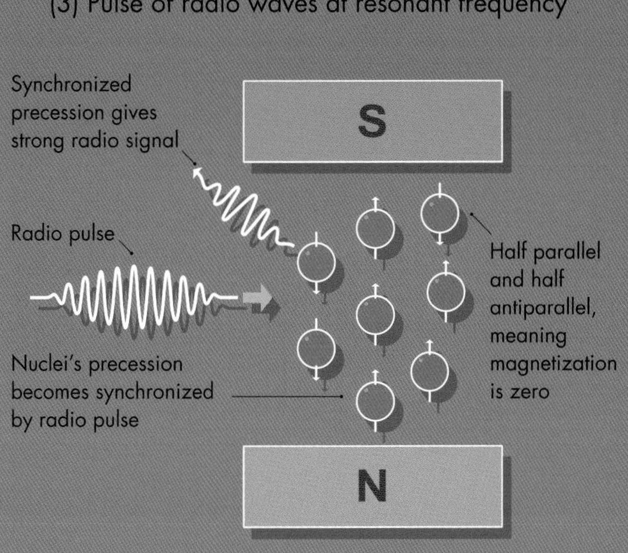

(4) Nuclei relax back to previous state

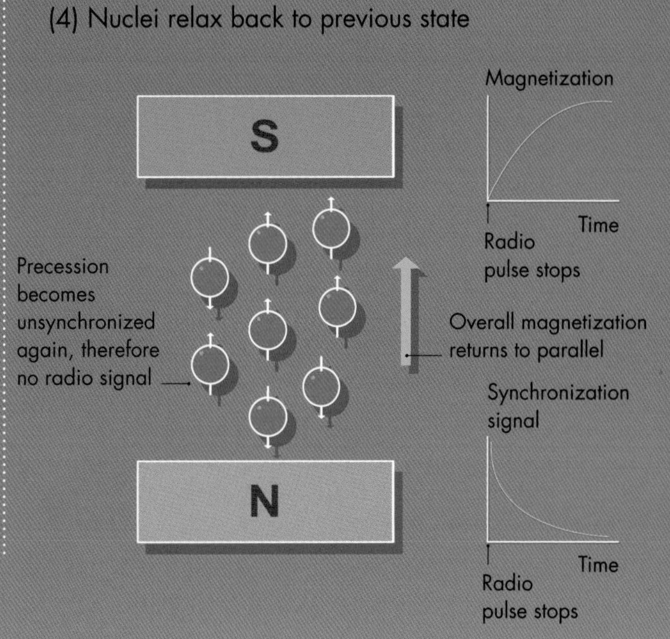

RADIOACTIVITY

In the twentieth century, scientists became aware of both the dangers and the potential uses of radioactive substances. In many applications, it is the particles and rays produced by radioactive substances that are of use. For the purpose of diagnosing and treating a range of diseases, for example, the interaction of the radiation with the body is of utmost importance. But some applications simply use the heat generated by nuclear disintegration.

A radioactive nuclide is any combination of protons and neutrons that is unstable. An unstable nucleus can "decay" to a more stable state by emitting alpha particles, beta particles, or gamma rays (see page 61). These emanations are called ionizing radiation, because they knock electrons from atoms, thereby creating ions.

RADIOACTIVITY IN MEDICINE

Some forms of radiation therapy, or radiotherapy, used to treat cancer, use radioactive substances. The ionizing radiation causes damage to the DNA in cells in and around a tumor; the cancerous cells cannot repair themselves, while cells in the normal tissue can. The therapeutic dose of radiation is delivered from within the body via medicines called radiopharmaceuticals. Radiotherapy can also be delivered by high-powered X-rays instead; these have a similar effect on DNA as the ionizing radiation from radioactive substances. Typically, radiopharmaceuticals contain radioactive nuclides that decay by alpha decay, because the emitted alpha particles are highly ionizing but have a short range, so they can have a localized effect.

Often, physicians inject or surgically place a small amount of the radiopharmaceutical inside or next to a tumor, a

IONIZATION IN THE HOME

Most domestic smoke detectors have a tiny sample of the radioactive nuclide americium-241. It produces a constant stream of alpha particles that ionize the air inside the detector. A small current flows through the ionized air. When smoke particles are present, they absorb the alpha particles, so the air is no longer ionized, and the current is interrupted, triggering the alarm. This is the opposite of how a Geiger-Müller radiation detector works (see page 60). Inset: The international hazard symbol for ionizing radiation.

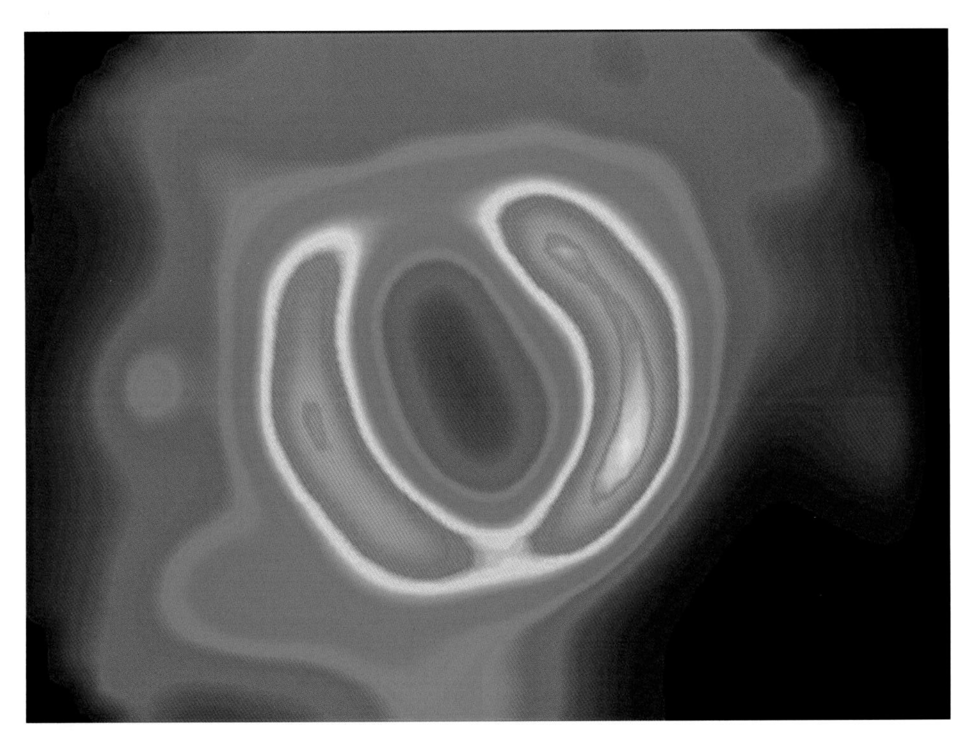

Scan showing the blood flow through the heart muscle of a patient following a heart attack. The radioactive nuclide technetium-99m was injected into the blood prior to the scan.

A CT scan that highlights small radioactive sources (iodine-125, seen as red in the image below) that have been surgically inserted into a patient's prostate gland in an attempt to treat a localized cancer.

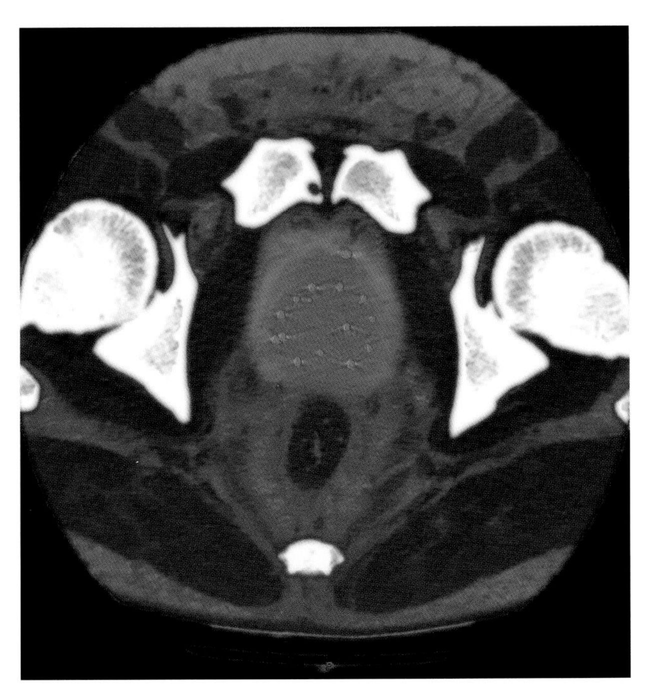

procedure called brachytherapy, which is commonly used to treat breast, prostate, and cervical cancers. But there are other ways in which the radiation can be delivered to the tumor. For example, skin patches containing phosphorus-32 can be used to treat skin cancer.

Many radiopharmaceuticals are used for aiding diagnosis, not treatment, typically by producing an image of blood flow or the shape and location of a tumor. The radioactive nuclides are attached to compounds that are taken up preferentially by certain tissues, acting as radioactive tracers. The medicines can be injected, taken as pills, or inhaled as a gas. In this case, the radioactive nuclide involved is normally an emitter of gamma rays, because these are less ionizing than alpha and beta particles, so escape the body largely without interacting with the living tissues. The gamma rays can then be detected and used to build up an image of the relevant area.

DECAY HEAT

If the radiation emitted by a radioactive substance is absorbed within the bulk of a material, instead of escaping, it will ultimately manifest as heat. Deep beneath your feet, natural radioactive isotopes of uranium, thorium, and other heavy elements produce enormous amounts of heat that keep the rocks of Earth's mantle molten, drive the movements of the tectonic plates, and are the ultimate source of geothermal energy. On a smaller scale, residual radioactivity in spent nuclear fuel continues to produce heat for years after a power station is decommissioned. Enough heat is generated within even a small block of plutonium-238 to make the block glow red-hot.

Space flight engineers use decay heat in two ways. Space probes that landed on the Moon and Mars carried radioisotope heating units (RHUs). These small devices contain a few ounces (tens of grams) of radioactive isotopes, normally polonium-210 or plutonium-238, which produce a few watts of heat—enough to protect the onboard electronics from extremely cold lunar or Martian nights. Many space probes rely on decay heat to provide electrical power, especially those venturing so far out in the solar system that solar panels cannot produce sufficient energy to power the spacecraft's electrical systems. These craft carry radioisotope thermionic generators (RTGs), in which an array of thermocouples generates a voltage. A thermocouple is a junction of two dissimilar metals, which produces a voltage when one metal is at a different temperature from the other. In an RTG, one side of the junction is exposed to the cold of space, while the other is close to the radioactive source (normally plutonium-238). A refrigerator-size RTG produces a few hundred watts, but that figure will decrease over time as the radioactivity diminishes.

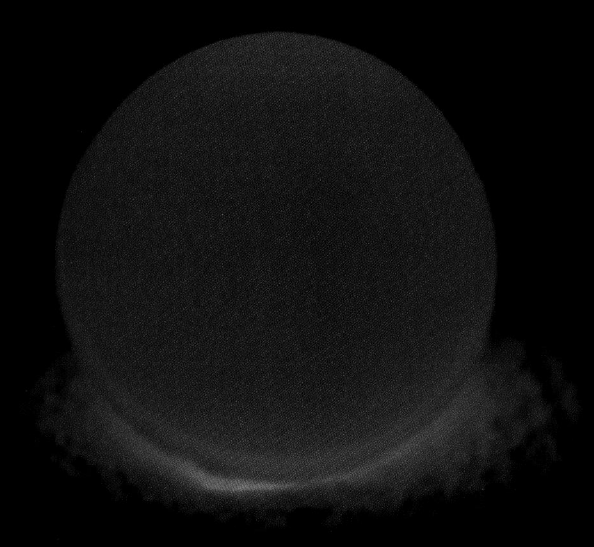

A sphere of plutonium naturally glows red-hot as a result of the heating effect of energy released by countless radioactive decay events inside.

An artist's impression of the New Horizons space probe (facing page) as it approached the dwarf planet Pluto (and its moon, Charon, behind). Like most space probes, New Horizons carries a radioisotope thermionic generator that generates electricity from the decay heat of the radioactive isotopes it contains. So far from the Sun, solar panels generate hardly any electrical power.

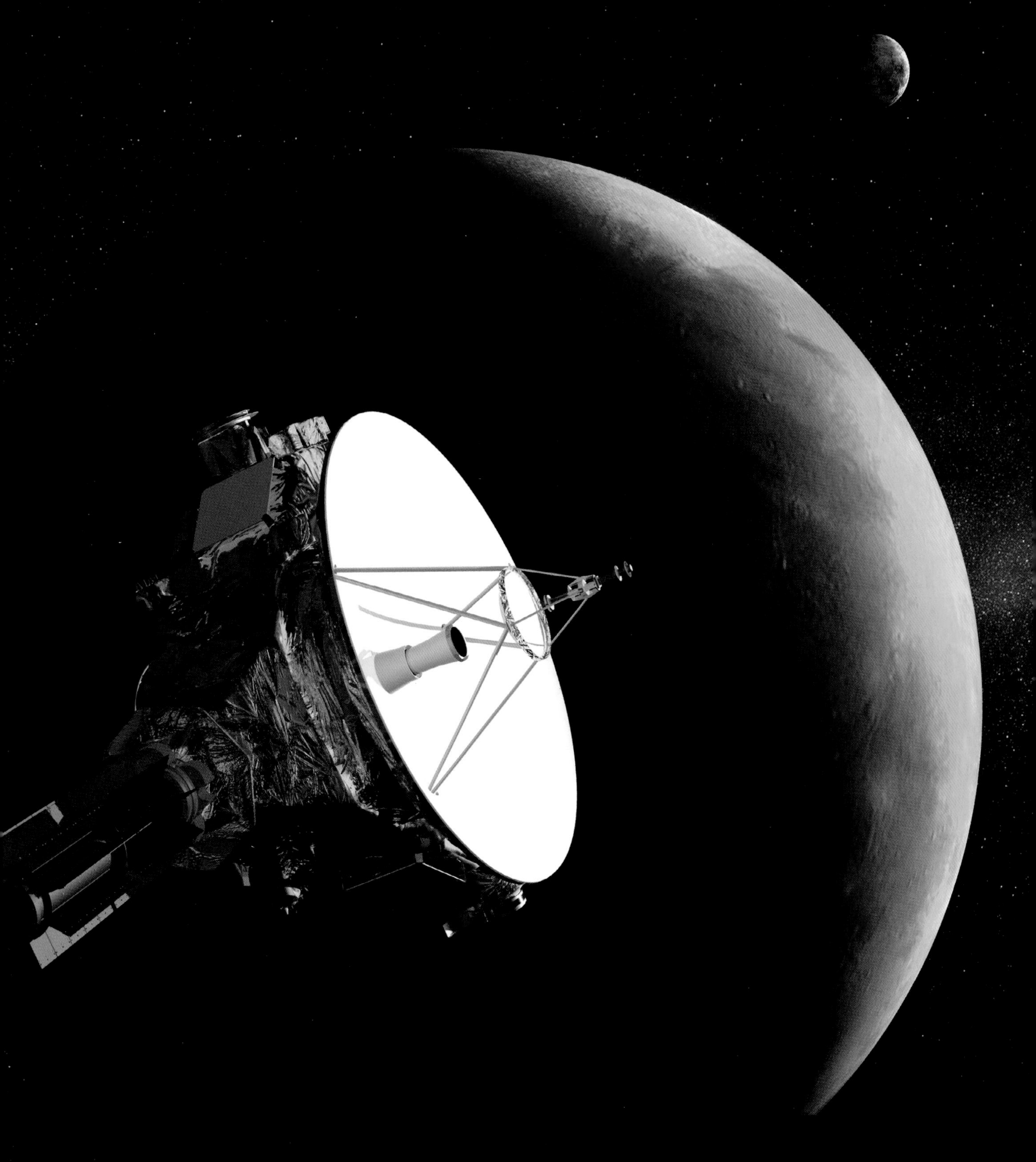

RADIOMETRIC DATING

Another application of radioactivity that is more relevant to scientific investigations than to everyday life is radiometric dating: the ability to estimate the ages of geological or archeological specimens. The more time passes, the more unstable atoms in a particular sample will decay. This makes it possible to work out the age of the sample by measuring the proportion of undecayed nuclides to decay products. The most common versions of radiometric dating are radiocarbon dating and potassium-argon dating.

Using an electric rasp, a scientist removes a sample from a fragment of fossilized human skull. The resulting powder will be tested for its carbon-14 content to give an estimate on how long ago the human died.

Radiocarbon dating is useful for measuring the age of long-dead biological materials. It is based on the fact that living things take in carbon while they are alive but not once they are dead. Plants take in carbon by photosynthesis, while animals absorb it by consuming plants that have been photosynthesizing. A small proportion of the carbon that a living thing will assimilate during its lifetime is the radioactive nuclide carbon-14. This is produced in the atmosphere when cosmic rays from space collide with atoms and produce free neutrons.

When these neutrons hit atoms of nitrogen-14 (7p, 7n), they produce carbon-14 (6p, 8n), plus a free proton. Carbon-14 is being produced at a constant rate in the atmosphere. Carbon-14 decays back into nitrogen-14 by beta decay (see page 60), with a half-life of about 5,700 years.

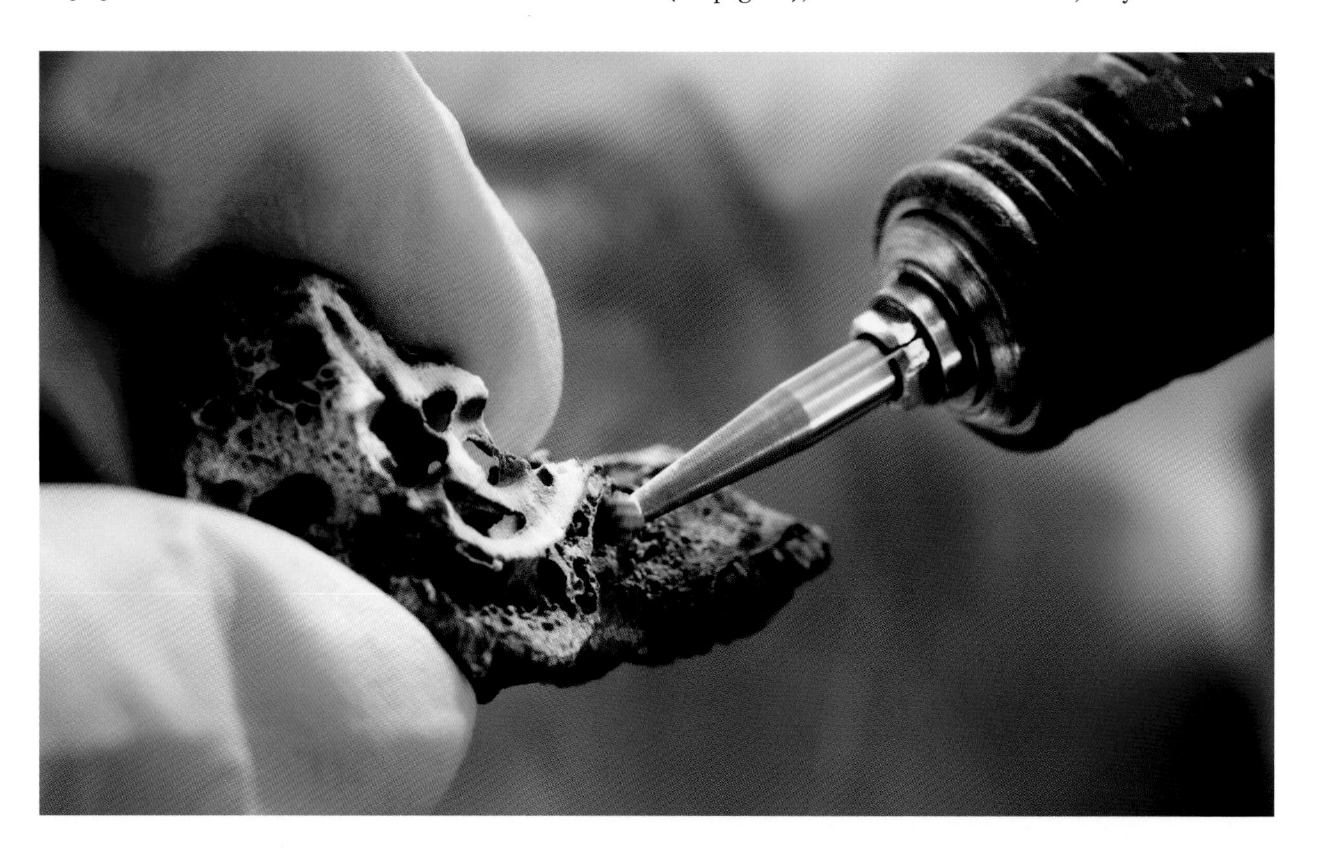

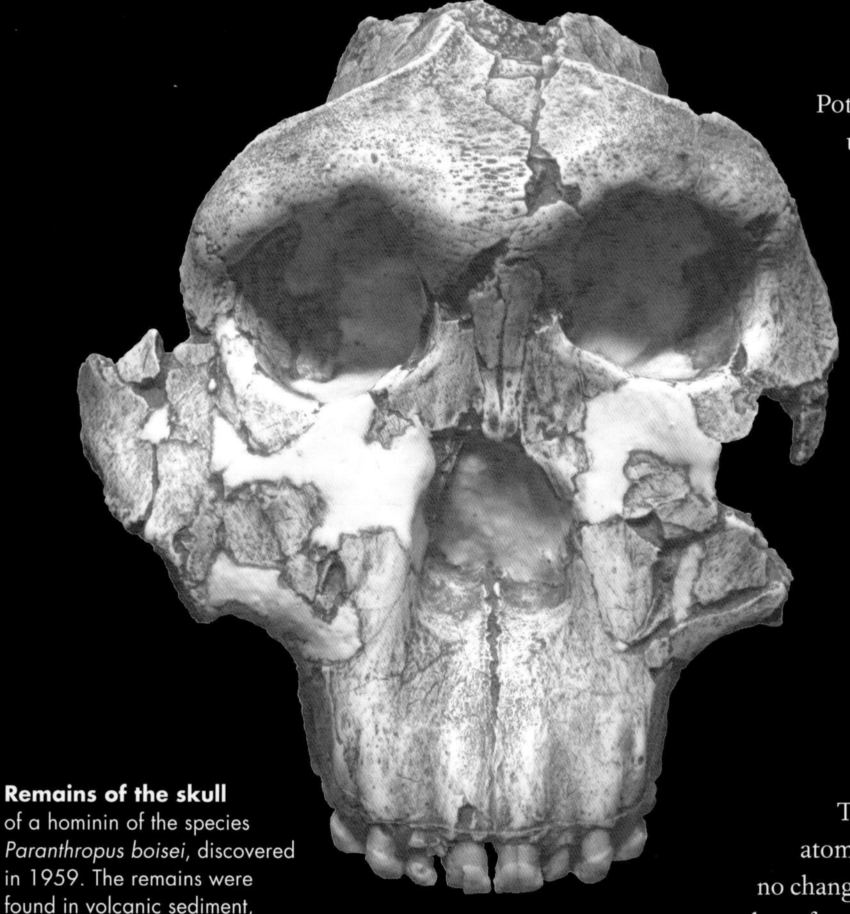

Remains of the skull of a hominin of the species *Paranthropus boisei*, discovered in 1959. The remains were found in volcanic sediment, which meant that scientists could use potassium-argon dating to work out its age (1.75 million years).

While a living thing is alive, it is constantly replenishing its complement of carbon-14, so that it has the same proportion of it as the atmosphere. But once the organism is dead, it amasses no more, and the carbon-14 decays. Measuring the ratio of undecayed carbon-14 still present to normal nonradioactive carbon-12 is indicative of how long ago the organism died. That measurement is normally made in a mass spectrometer (see page 69). With radiocarbon dating, it is possible to estimate the date of organic materials used in buildings and clothing during prehistory, as well as bone and other fragments of once-living specimens.

Potassium-argon dating makes use of the fact that the naturally occurring nuclide potassium-40 is unstable. About nine out of ten potassium-40 nuclei decay by beta decay, resulting in calcium-40—this is stable and commonplace, and is of no use in radiometric dating. However, about one out of ten potassium-40 nuclei decay by a different process, called electron capture. As the name suggests, electron capture involves a nucleus absorbing an electron; the electron combines with a proton, changing it into a neutron. This has the effect of reducing the atomic number by one, but makes no change to the atomic mass (the total number of protons and neutrons). Potassium has an atomic number of 19, and the element with the atomic number 18 (one less) is argon. And so, when a potassium-40 nucleus captures an electron, it becomes a nucleus of argon-40. Argon is a noble gas (see page 80) so does not react chemically with any other elements. If produced by radioactive decay of potassium atoms in a rock, it becomes trapped in the rock's crystal structure. Potassium-argon dating is useful for igneous rocks, produced from solidified lava. Before the rock becomes solid, any argon already present can escape, but afterward, any new argon produced becomes trapped. This makes it possible to work out how long ago the rock formed. Once again, mass spectrometry is used to measure the ratio of argon-40 to potassium-40 in a sample of rock.

NUCLEAR REACTIONS

Most of the radioactive nuclides used in medicine and in space flight are produced in nuclear reactors. All commercial nuclear reactors involve nuclear fission, in which large, heavy nuclei break apart into smaller fragments, releasing large amounts of energy and radioactive products. But there is another type of nuclear reaction, nuclear fusion, which holds promise for near limitless energy production in the future.

NUCLEAR FISSION

Research into the structure and behavior of the atomic nucleus began in earnest in the 1920s. By the late 1930s, physicists understood how some large, unstable nuclei could break into sizable fragments—or "fission"—instead of simply release alpha particles or beta particles, or lose energy as gamma rays. Fission can take place spontaneously, but it can also be encouraged to happen by bombarding unstable nuclei with neutrons. Because free neutrons are released when a large, unstable nucleus fissions, this process can become a chain reaction. In this process, the newly released neutrons cause fission in other nearby nuclei, which then go on to release more neutrons when they break up. A chain reaction can start and be sustained only if enough unstable nuclei are present. There is, therefore, a critical mass below which a chain reaction does not occur.

The smaller daughter nuclei produced by nuclear fission have less energy than the unstable nucleus from which they form. So, in a lump of fissile material with critical mass in which a chain reaction is occurring, large amounts of energy are released as the recoil of the nuclear fragments and the motion of the neutrons. Enough energy is released, quickly enough, to boil water in the reactor core of a nuclear power station (the steam then powering electrical generators) or to cause a devastating explosion when an atomic bomb is detonated.

The most common fissile nuclide used in nuclear reactors is uranium-235, and in atomic bombs, uranium-235 or plutonium-239. However, some nuclear reactors use thorium-232. There is a more plentiful supply of thorium in Earth's crust than there is uranium, and a thorium reactor produces less long-lived radioactive waste than a uranium reactor. However, there are technical difficulties in using thorium that prevent thorium reactors from becoming the predominant type of reactor.

The core of a nuclear reactor during maintenance. In operation, this core is filled with water contained at high pressure so it does not boil, but it can transfer heat to water that can boil to drive turbine generators.

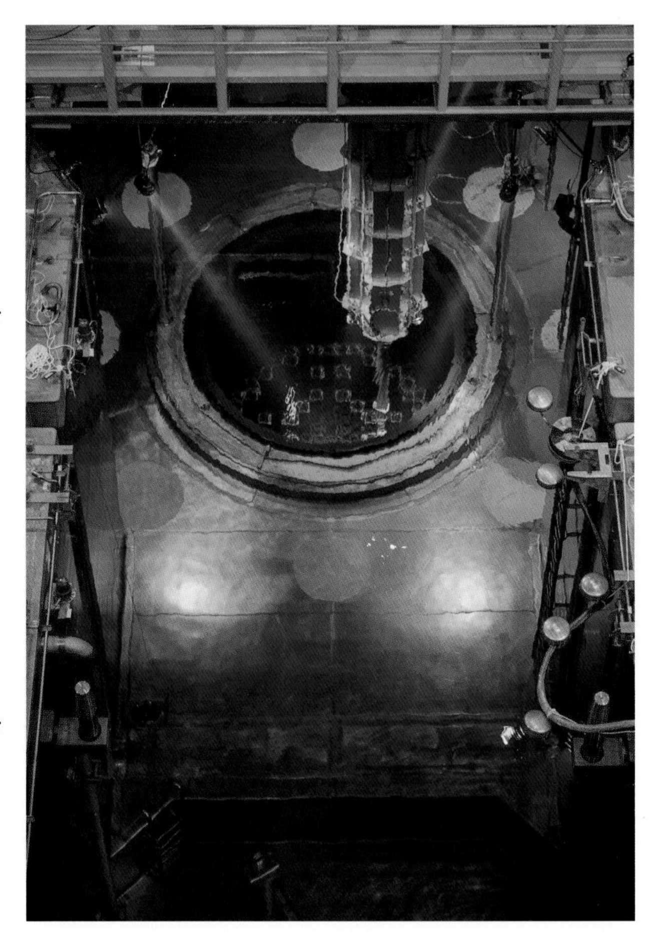

FISSION CHAIN REACTION

When certain unstable nuclides (here, uranium-235) absorb a neutron, they split into two smaller-size fragments—two smaller nuclides. In the case of uranium-235, these are typically krypton-92 and barium-141. Energy is released, because the binding energy (see page 41) of the smaller nuclei is less than the binding energy of the large one. Note how 92 and 141 add up to 233. Along with the original neutron, there were a total of 236 nucleons present before the fission. Three neutrons are therefore present, and if there are other uranium-235 atoms nearby, these neutrons will cause more fission events. It is easy to see how this process can quickly become a chain reaction, releasing huge amounts of energy rapidly. In a nuclear reactor, the chain reaction is controlled by inserting rods made of materials that absorb some of the free neutrons. In an atomic bomb (see page 162), the chain reaction continues unabated, with terrible destructive results.

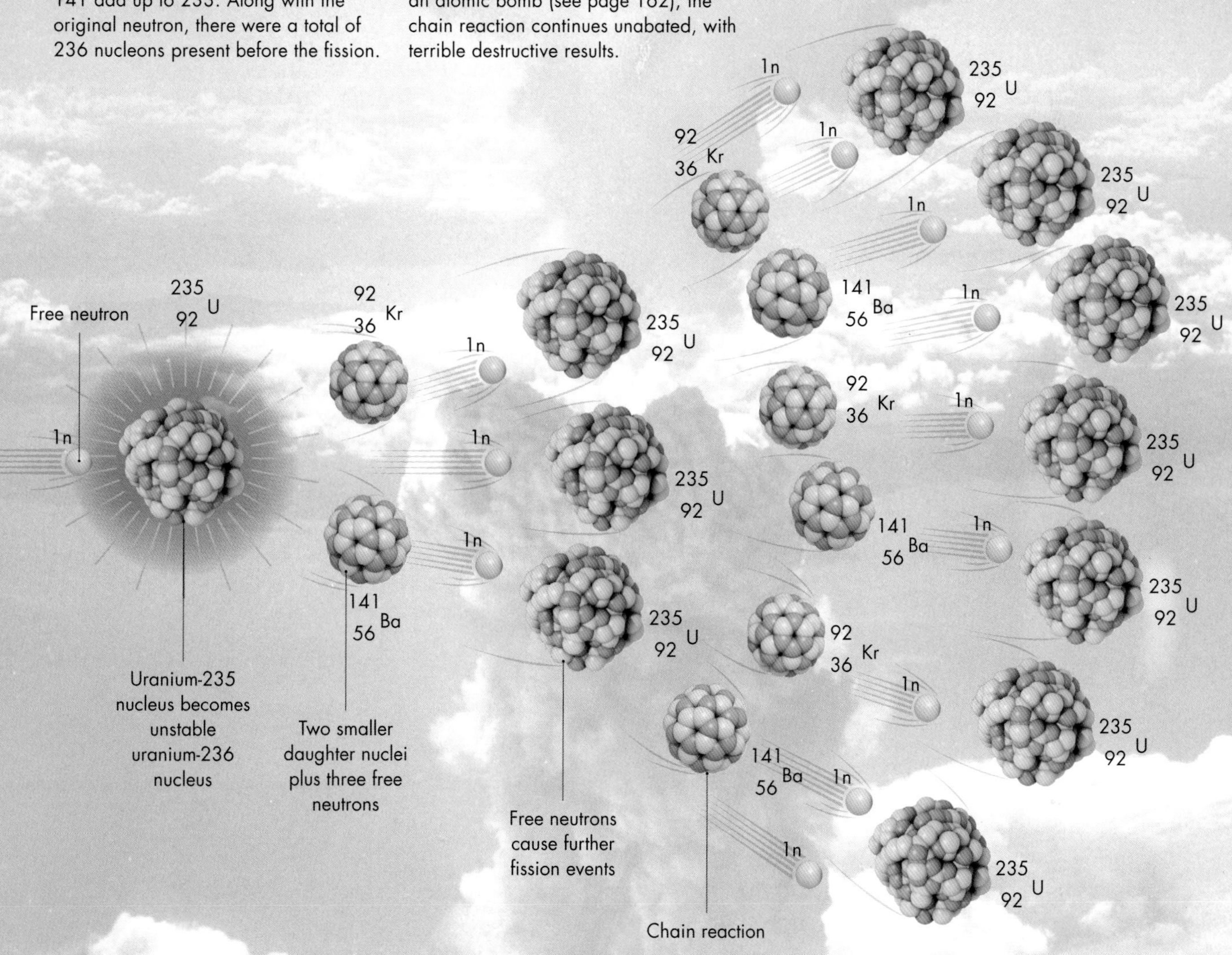

1

2

3

NUCLEAR FUSION

Large amounts of energy are also released during nuclear fusion—the nuclear reaction that powers the Sun and other stars (see page 72). In nuclear fusion, two small nuclei (typically isotopes of hydrogen) are forced together under conditions of extreme temperature and pressure to form a new, larger nucleus. The new nucleus has less energy than the two smaller ones from which it formed, and, once again, the excess energy is released as heat. This process can also be used in weapons, such as the hydrogen bomb, also called a thermonuclear device. However, fusion researchers hope that one day controlled fusion may generate cheap and plentiful electricity. A fusion reactor creates virtually no radioactive waste and uses as its fuel hydrogen—the most plentiful element in the universe.

Nuclear fusion only happens at extremely high temperatures, when the nuclei are moving rapidly enough to collide with sufficient momentum that fusion becomes a possibility. Temperatures of tens of millions of degrees are required—and one problem with fusion power technology is how physically to contain such an incredibly hot substance. At such high temperatures, the hydrogen is completely ionized, existing as a plasma—and fortunately, there are ways to confine a plasma without it touching the walls of whatever contains it. In particular, plasmas can be confined and controlled using strong magnetic fields. The usual way of doing this is in a toroidal (doughnut-shape) chamber called a tokamak. Nuclear fusion has been initiated many times in experimental reactors, but so far the energy used to start the reaction has always been far greater than the energy harvested from it.

The first thermonuclear explosion (nuclear fusion) in a test firing (1) of a hydrogen bomb codenamed Ivy Mike, at the Enewetak Atoll, near the Marshall Islands in the Pacific, in 1952. Hydrogen plasma (2) at a temperature of several millions of degrees inside a tokamak reactor and (3) inside a tokamak when not in operation.

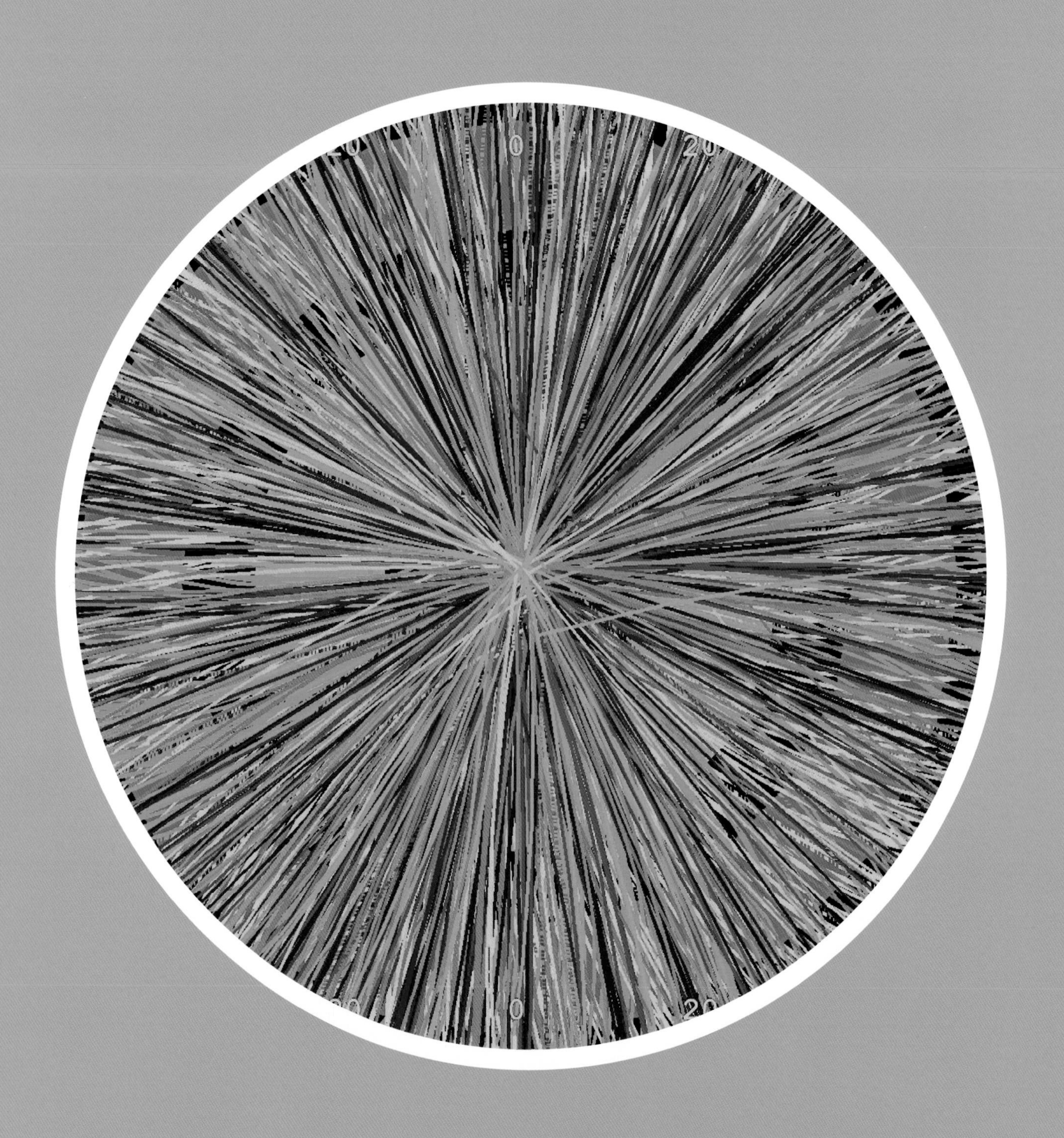

CHAPTER 7 THE END OF ATOMISM?

The overriding tenet of atomism is that matter is made of tiny, indivisible particles: atoms. In the twentieth century, physicists realized that atoms have internal structure—that they are themselves made of smaller parts. The nucleus is made of protons and neutrons, which are themselves made of quarks. The scientists went on to discover a host of other subatomic particles, raising the question: what are the real atoms in the world? The truly fundamental, uncuttable parts of matter? What is the world really made of? The answer, it seems, is fields.

The tracks of thousands of particles produced by the high-energy collision of lead ions at CERN, near Geneva, Switzerland. Experiments like these have allowed physicists to probe matter at its most fundamental level and determine the interactions of the many kinds of particles with sizes or masses at or below the atomic scale.

THE SEARCH FOR THE ULTIMATE ATOMS

The original meaning of "atom" suggested that it should be "fundamental," or "elementary": an indivisible solid particle made of nothing else, with no inner structure. But modern physics has discovered that atoms, tiny though they are, are not nature's ultimate building blocks. Instead, they are made of even smaller objects—and there exists a plethora of other tiny particles. Just as the periodic table finds order in the chemical elements, these particles are organized under a theory called the Standard Model.

The periodic table was an attempt to find order in the bewildering diversity of matter in the world around us. It was conceived before physicists knew of the existence of protons and neutrons, or even of electrons—but it turns out that the order in the table is a consequence of atomic structure based on the unintuitive but relatively simple rules of quantum mechanics. Each place in the periodic table is occupied by an element whose atoms have a certain number of protons and electrons, and the chemical properties of each element are determined by the way in which its electrons are organized around the nucleus. As such, physicists and chemists were not disappointed by the fact that atoms have an inner structure, because that inner structure helps to explain the patterns in the periodic table (see page 78). But the quest to find the true atoms—fundamental particles with no internal structure— continued.

MORE PARTICLES

After the discovery of the neutron, the picture of an atom as a nucleus made of protons and neutrons surrounded by a cloud of electrons persisted. To start with, atomic physicists could console themselves with the notion that while atoms were clearly not fundamental, at least protons, neutrons, and electrons were. And yet, through the twentieth century, a combination of theory and experiment uncovered a plethora of particles not previously dreamed of and eventually led scientists to realize that even protons and neutrons are not fundamental. They shook the neat idea of a world made simply of protons, neutrons, and electrons to its core.

First came British physicist Paul Dirac. In the late 1920s, Dirac found a way to combine quantum theory with relativity. He unwittingly discovered that the electron has a doppelgänger, or double, identical in everything except electric charge. The antielectron, or positron, has the same mass as the electron, but it carries a positively charge, instead of a negative one. The inescapable conclusion was that every particle has an associated

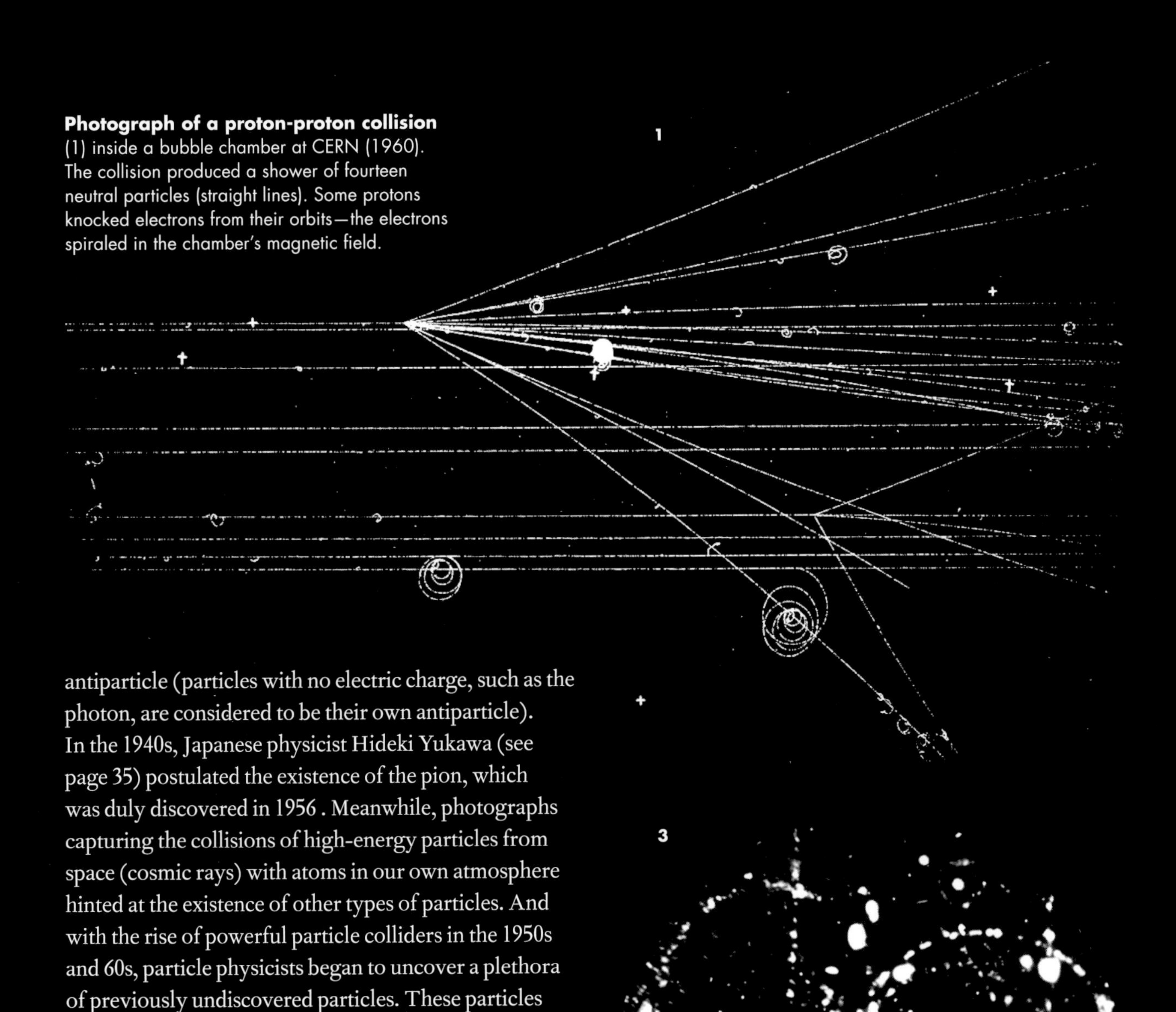

Photograph of a proton-proton collision (1) inside a bubble chamber at CERN (1960). The collision produced a shower of fourteen neutral particles (straight lines). Some protons knocked electrons from their orbits—the electrons spiraled in the chamber's magnetic field.

antiparticle (particles with no electric charge, such as the photon, are considered to be their own antiparticle). In the 1940s, Japanese physicist Hideki Yukawa (see page 35) postulated the existence of the pion, which was duly discovered in 1956 . Meanwhile, photographs capturing the collisions of high-energy particles from space (cosmic rays) with atoms in our own atmosphere hinted at the existence of other types of particles. And with the rise of powerful particle colliders in the 1950s and 60s, particle physicists began to uncover a plethora of previously undiscovered particles. These particles were not part of an atom, and so did not make up ordinary matter as we know it—but what were they?

False color image (2) from an airborne photographic plate (1950) that recorded the collision of a cosmic ray particle (red) colliding with a nucleus in the photographic emulsion, producing pions (yellow), a fluorine nucleus (green),and other nuclear fragments (blue). (3) The first photograph (right) of a cosmic ray collision in a cloud chamber (1927). Note how charged particles bend in the magnetic field inside the chamber.

In 1952, American physicist Donald Glaser invented the bubble chamber, a tank of liquid (normally hydrogen) in which electrically-charged particles would leave tracks of tiny bubbles. Magnetic and electric fields inside the tank caused the particles to follow curved paths that depended upon the particles' mass and charge. This device helped particle physicists to discover a slew of new particles that were produced by collisions between other particles in particle accelerators. Each particle had a different combination of charge and mass, and interacted with some particles but not others. Mysteriously, particles were also able to come mysteriously in and out of existence.

In the 1960s, American physicist Murray Gell-Mann and Russian-American physicist George Zweig tried to make sense of the many new particles by suggesting

particles that Gell-Mann named "quarks." Their theory proposed that some particles were made of two quarks each, while protons and neutrons (and antiprotons and antineutrons) were made of three quarks each. The scientific community was dubious at first, but experiments in the 1970s proved the theory correct and confirmed the existence of quarks. Any particle consisting of quarks is a hadron, and any hadron made of two quarks is a meson, while any three-quark hadron is a baryon. At first, only two types of quark were considered: the "up" quark and the "down" quark (for example, a proton consists of two "up"s and one "down"). But it soon became clear that there must be others—equivalent quark pairs with higher masses, which were given the epithets "charm" and "strange," and "top" and "bottom." And, of course, each one has

False color proton-proton collision (1) in a CERN bubble chamber. Charged particles follow spiral paths. In the false-color particle collision (2), the particle tracks are streams of tiny bubbles in neon and hydrogen mixture inside the bubble chamber at CERN. A gamma ray photon spontaneously creates an electron-positron pair (3). The new particles spiral away from each other, producing more photons that create more particle-antiparticle pairs. In a bubble chamber (4), a kaon (k-meson) collides with a hydrogen nucleus (proton) at the bottom of the image.

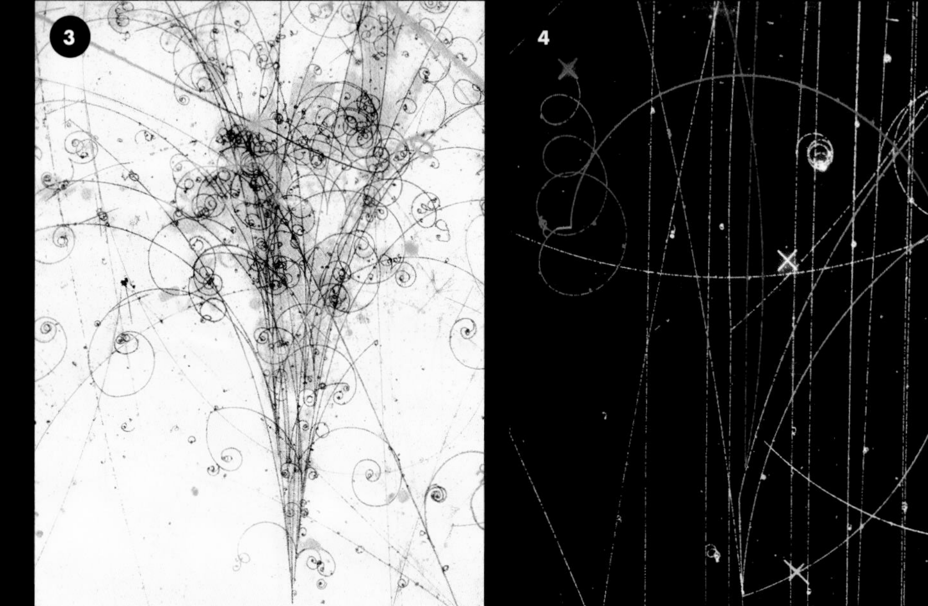

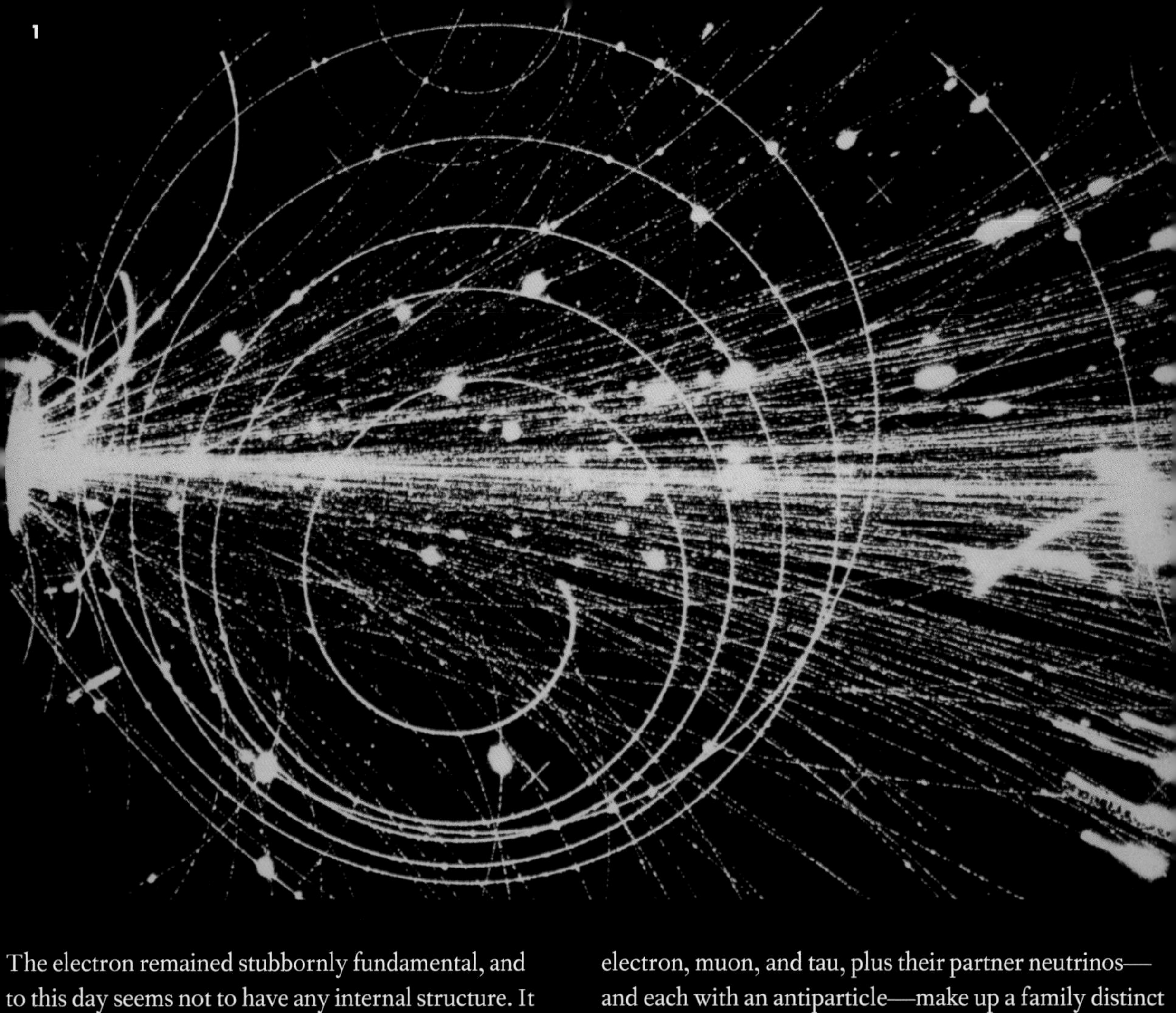

The electron remained stubbornly fundamental, and to this day seems not to have any internal structure. It does, however, have cousins, in the shape of the muon and the tau (and, again, their antiparticles, the antimuon and the antitau). These particles are alike in charge and general behavior, but they have more mass than the electron—and, unlike electrons, they are not constituents of ordinary matter. Yet another particle, the neutrino, is closely associated with the electron. Its existence was postulated in 1930 to help explain the mysterious loss of energy that occurs during beta decay (during which process a nucleus ejects an electron). The neutrino was discovered in the 1950s—and it, too, has more massive relations, the muon neutrino and tau neutrino. The electron, muon, and tau, plus their partner neutrinos—and each with an antiparticle—make up a family distinct from the hadrons, called the leptons.

By the 1970s, it seemed that matter was made of the (quark-based) hadrons plus the leptons. Alongside these particles were others that carried, or mediated, the forces between the particles of matter. These force-carrying particles are "gauge bosons": the photon (which carries the electromagnetic force), the gluon (which carries the strong force between quarks), and th W and Z bosons (which carry the weak nuclear force involved in nuclear decay, more properly called the weak interaction).

An oxygen ion collides with a lead nucleus at the left of the image (1), producing a shower of particles. The large spiral is the track left by a low-energy electron. (2) The decay of a positive kaon (K+) captured in a bubble chamber. The products of the decay speed and spiral away and some cause secondary collisions. (3) The discovery of "neutral currents," one aspect of the weak interaction (1973). A neutrino, unseen, knocks an electron from its atom (seen bottom right). The electron shoots off to the left, creating electron-positron pairs.

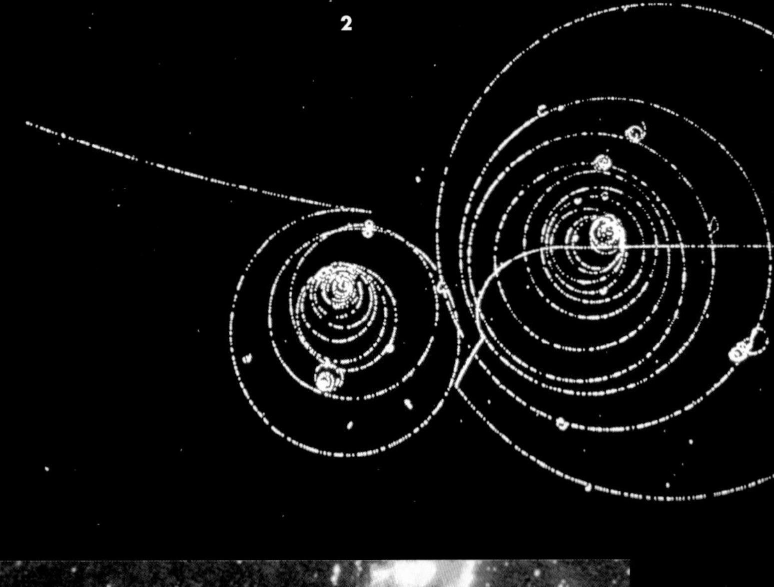

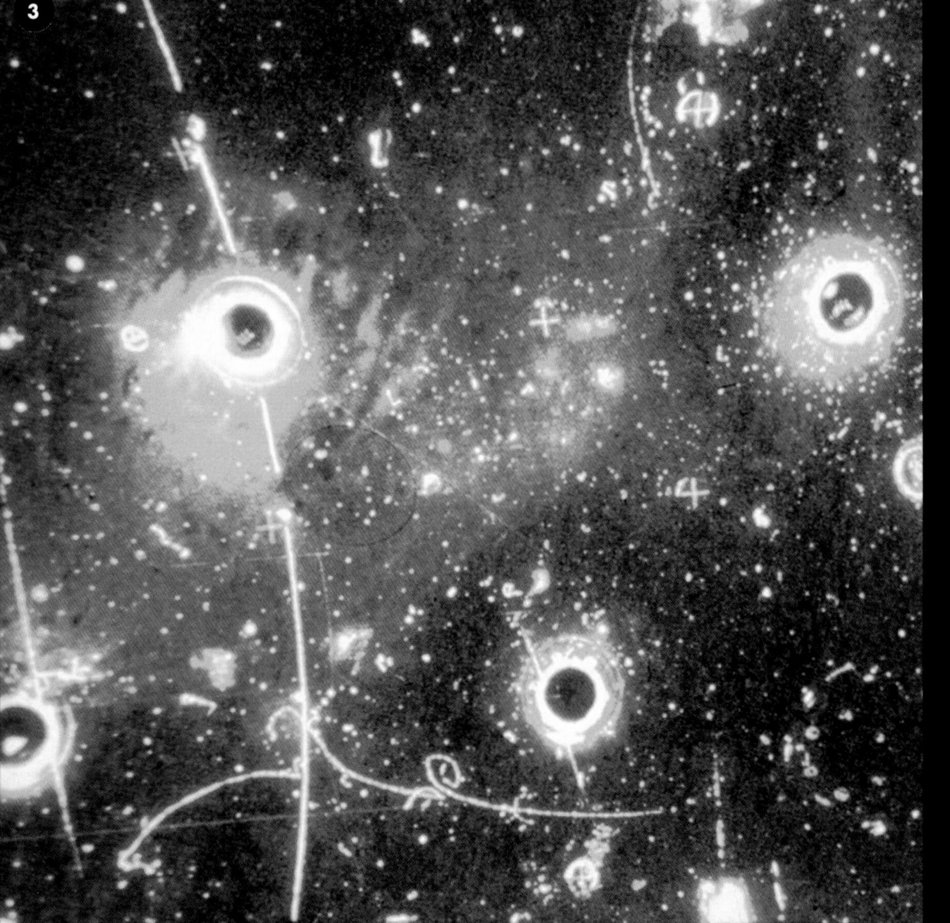

To make things more complicated still, there exists the quantum property called spin—something all elementary particles possess, and which makes charged particles into tiny magnets (see page 49). Some particles have half-integer spin (−3/2, −1/2, 1/2, 3/2, and so on); these are known as fermions, and only one fermion can exist in a particular quantum state. All leptons (including electrons) are fermions, as are all quarks. The force-carrying particles are bosons, particles with integer spin (−1, 0, 1, and so on). Any number of bosons can have exactly the same quantum state. Composite particles can be fermions or bosons, depending upon the total spin of the particles of which they are made. Some atoms, for example, can be bosons, and this is why it is possible for them to become Bose-Einstein condensates (many atoms occupying the same state, as a "superatom," see page 132) at very low temperatures.

With so many different particles, and so many different ways of categorizing them, theoretical physicists sorely needed a way to make sense of what physicists began to refer to as the "particle zoo." They needed something that could do for the world of elementary particles what the periodic table did for atoms and elements.

The tracks of a shower of particles (1) created by the collision of an extremely fast moving proton with a lead ion, at CERN in 2012. On the facing page is a screen grab (2) of the results of two proton-proton collisions in the CMS (Compact Muon Solenoid) detector in the Large Hadron Collider at CERN (2012). Below it (3) is a shower of spiraling charged particles produced after the collision of high-energy gold ions at the Relativistic Heavy Ion Collider at the Brookhaven National Laboratory, at Upton, New York.

1

2

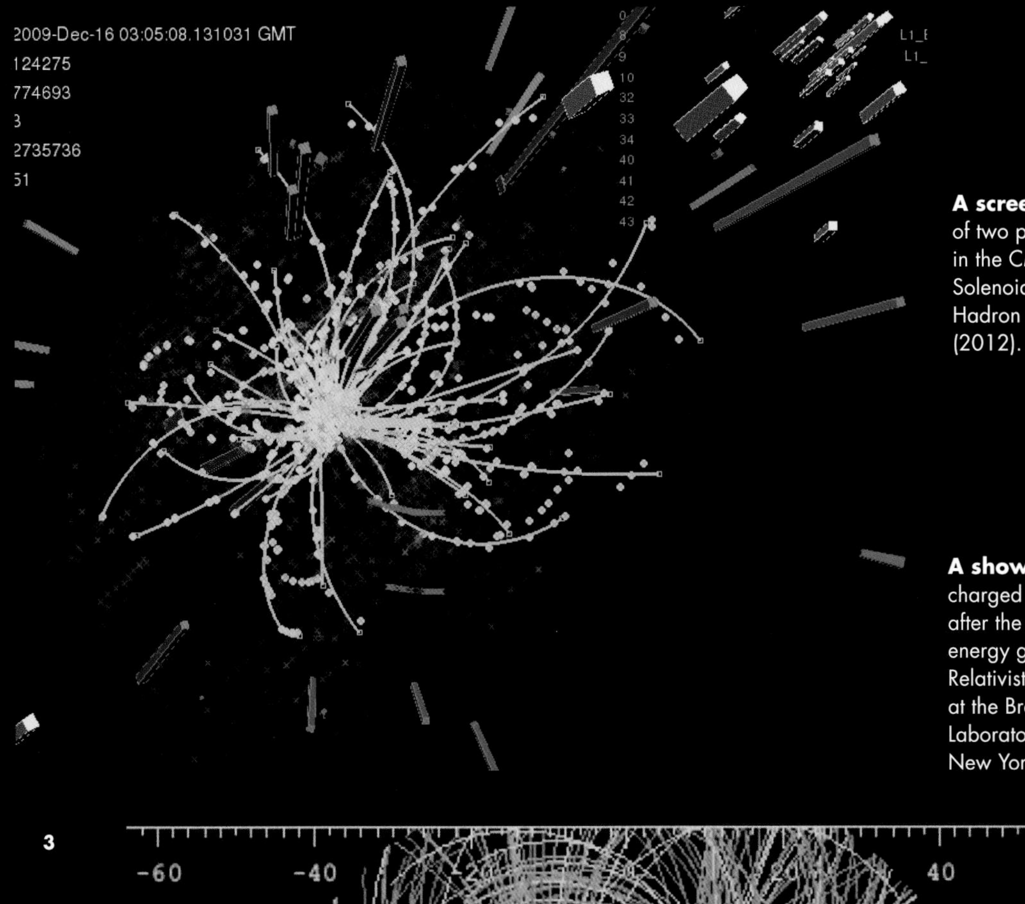

A screen grab of the results of two proton-proton collisions in the CMS (Compact Muon Solenoid) detector in the Large Hadron Collider at CERN (2012).

A shower of spiraling charged particles produced after the collision of high-energy gold ions at the Relativistic Heavy Ion Collider at the Brookhaven National Laboratory, at Upton, New York.

3

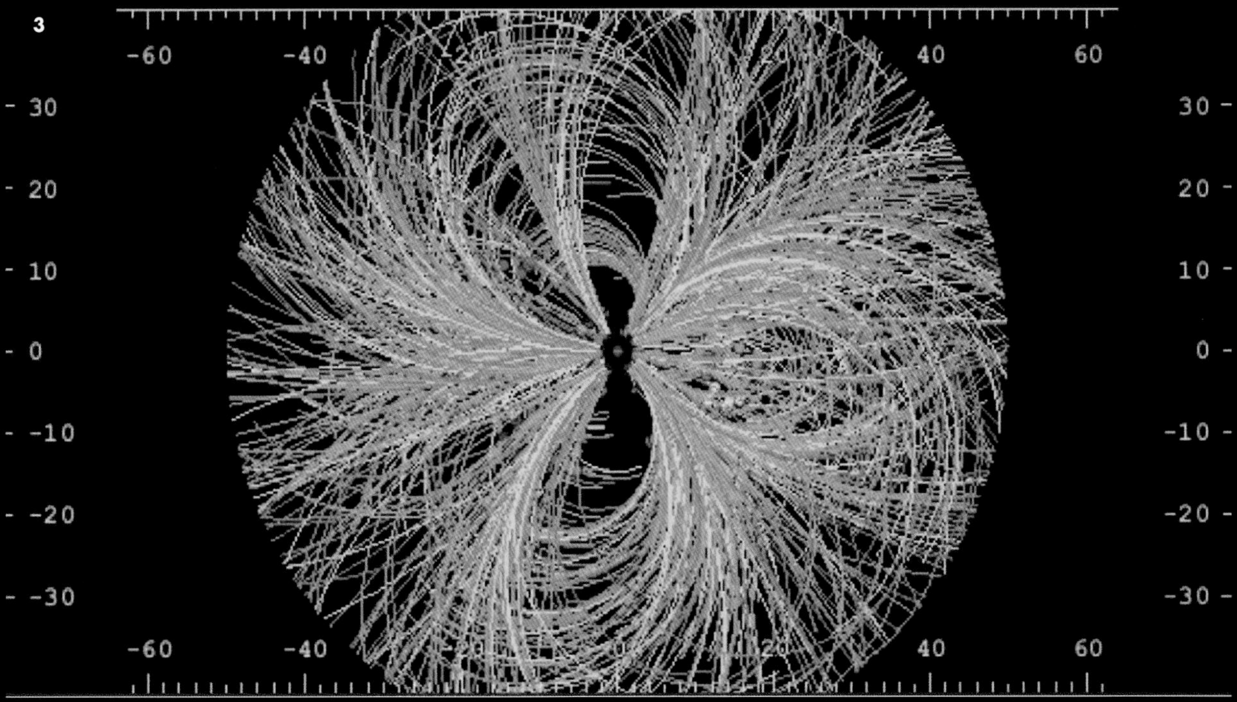

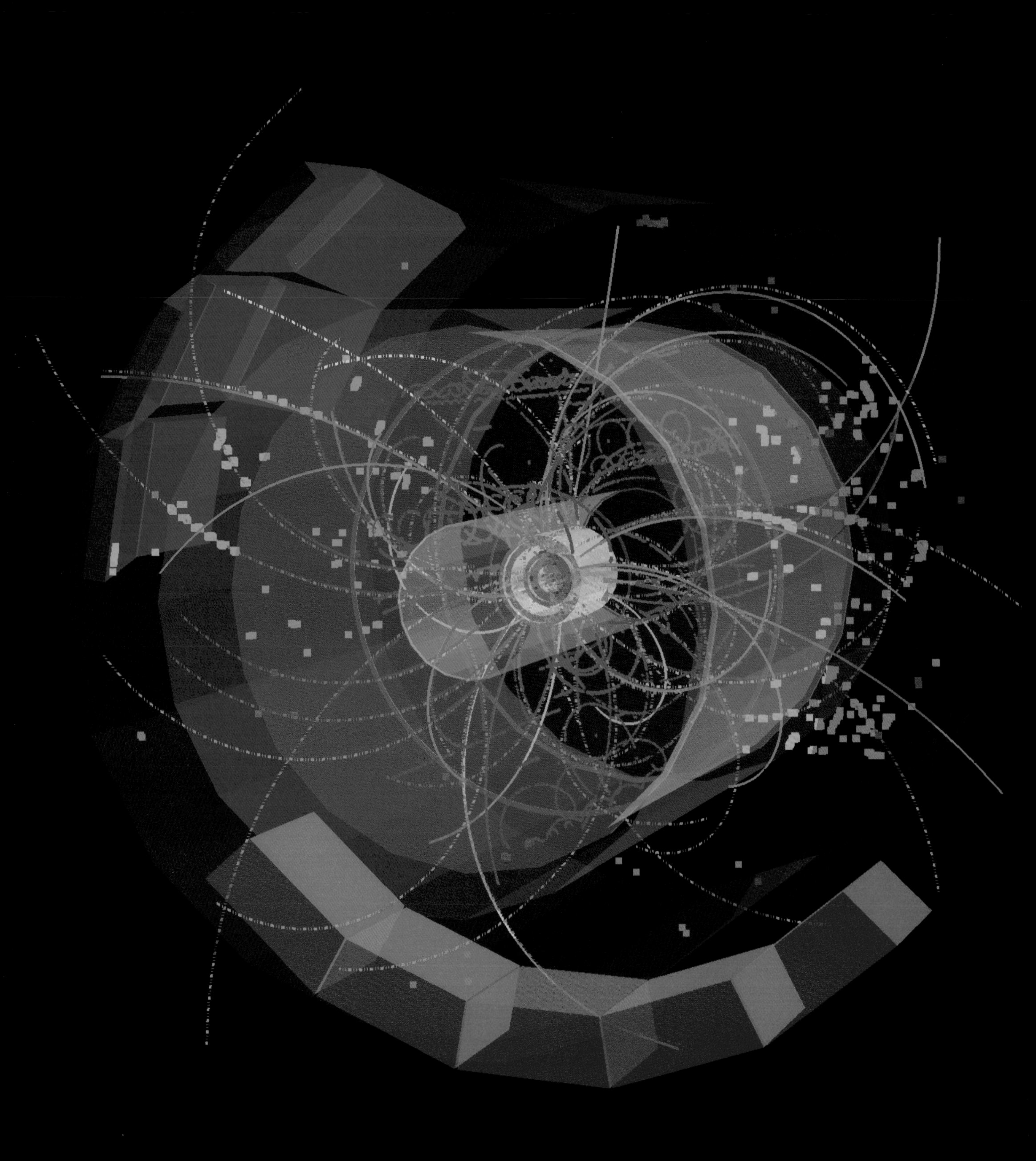

STANDARD MODEL

The scientists were aware that various interactions of the matter particles (hadrons and leptons), and the force-carrying particles (the gauge bosons), seemed to follow patterns. There are in fact four kinds of such interactions: gravity, electromagnetism, the strong nuclear force, and the weak interaction. Gravity remains stubbornly different from the other three types of interaction. It is beautifully explained by Einstein's theory of general relativity, and although there may still exist a particle that carries the force of gravity—the graviton—that particle has so far remained elusive. However, particle physicists have managed to codify the other three fundamental interactions. They have found order in what kinds of particle interact with what other, and how properties, such as charge and energy, must always be conserved in each interaction.

The result of this codification is an exquisite set of rules called the Standard Model, which can be summarized in a table showing the families of particles and how they interact. Order had been restored once again, and the members of the particle zoo tamed. The hundreds of particles that had been discovered can be accurately described using only twelve particles (six leptons and six quarks), plus four force-carrying particles (the gauge bosons).

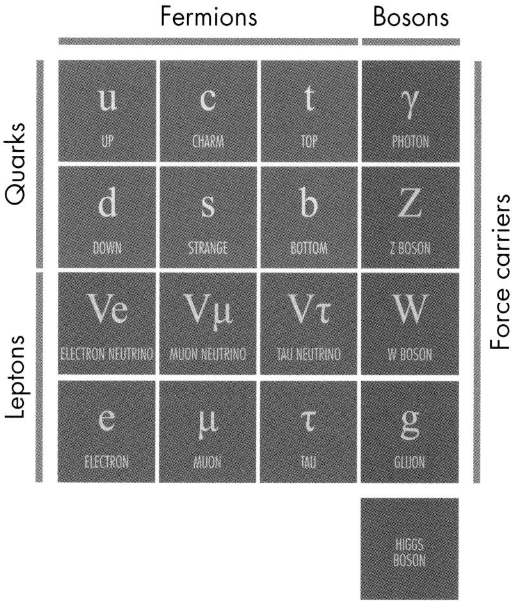

Computer reconstruction of a particle shower created by a collision of lead ions inside the ALICE detector at CERN.

There is an extra particle—another boson (but not a force-carrying gauge boson)—that was necessary to complete the Standard Model. This is the Higgs boson, and its celebrated discovery in 2012 in the Large Hadron Collider (LHC)—at the CERN research facility on the border between France and Switzerland—was the latest in a series of experimental results to support the efficacy of the Standard Model. Despite earning itself the epithet "the God particle," the Higgs boson itself is not particularly important, but its production at CERN confirmed the existence of an entity called the Higgs field. The Higgs field helps to explain various key phenomena in the Standard Model, including the origin of the mass of fundamental particles. Fields such as the Higgs field are a fundamental underpinning of the Standard Model.

The fundamental particles of the Standard Model. Fundamental fermions are matter particles; fundamental bosons are force-carrying particles (except the Higgs boson). Source: AAAS

	Fermions			Bosons	
Quarks	u UP	c CHARM	t TOP	γ PHOTON	Force carriers
Quarks	d DOWN	s STRANGE	b BOTTOM	Z Z BOSON	Force carriers
Leptons	Ve ELECTRON NEUTRINO	Vμ MUON NEUTRINO	Vτ TAU NEUTRINO	W W BOSON	Force carriers
Leptons	e ELECTRON	μ MUON	τ TAU	g GLUON	Force carriers
				HIGGS BOSON	

QUANTUM FIELDS

Underlying the Standard Model of particles and their interactions is the notion that particles are not solid objects; instead, they are disturbances in fields that permeate all of space. These fields are mathematical entities that help to describe such interactions, and they provide a crucial link between the wavelike and particle-like properties of what we normally think of as particles. In other words, they describe those particles in terms of quantum theory.

WHAT IS A FIELD?

The notion of fields dates back to the 1840s and was the brainchild of British scientist Michael Faraday. Throughout the 1820s and 30s, Faraday had been experimenting with electricity and magnetism. He realized that magnets and electric charges exert forces along definite lines that seem to permeate empty space. The strength of a field is different at different points in space; in other words, a magnet placed in a magnetic field will experience stronger or weaker forces at different locations within the field. In 1846, Faraday suggested that light might be a wave motion, a disturbance, within the electric and magnetic fields, an idea that James Clerk Maxwell confirmed mathematically twenty years later (see page 28). The fact that waves can propagate through fields is crucial—and became especially so when quantum theory exposed the wavelike nature of particles in the 1920s (see page 32).

The electric and magnetic fields exist as a single entity: the electromagnetic field. When physicists first applied quantum mechanics to understanding the atom, in the 1920s, they treated the energy states of the electron as

FEYNMAN DIAGRAMS

In 1948, American physicist Richard Feynman set out to simplify the complicated mathematical descriptions of the interactions of electrons and photons in QED, by representing them with simple diagrams. In each diagram, matter particles and force-carrying gauge bosons are shown as straight or wiggly lines. A composite particle, such as a proton, is shown as a collection of lines, each one representing a quark. There are many possible ways in which any interaction can take place.

When two electrons (e^-) repel one another, for example (right), they do so by exchanging virtual photons (γ). An electron can also exchange virtual photons with itself—or a virtual photon can spontaneously become a virtual electron-positron pair en route. Each diagram represents a possible mechanism behind the interaction, and can help to estimate the probability that mechanism will occur. Taken together, the diagrams suggest the overall probability that an interaction will take place.

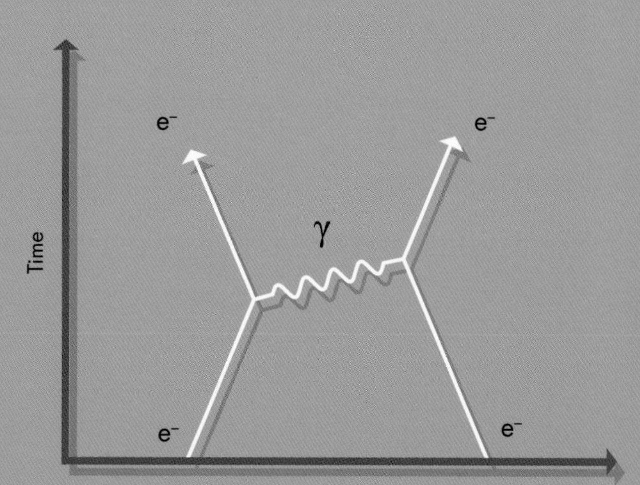

quantized (only certain energies are allowed), but did not really consider how the electromagnetic field itself could be quantized. Several physicists later tried to address this shortcoming and formulated a theory in which photons are disturbances, or excited quantizations, of the electromagnetic field. The study of the interaction of the electromagnetic field with charged particles is called electrodynamics, and the emerging theory that addressed the quantization of the electromagnetic field became known as quantum electrodynamics (QED).

Michael Faraday, the scientist whose experiments contributed greatly to our understanding of electromagnetism. He used the word "field" for the first time in a diary entry dated November 7, 1845.

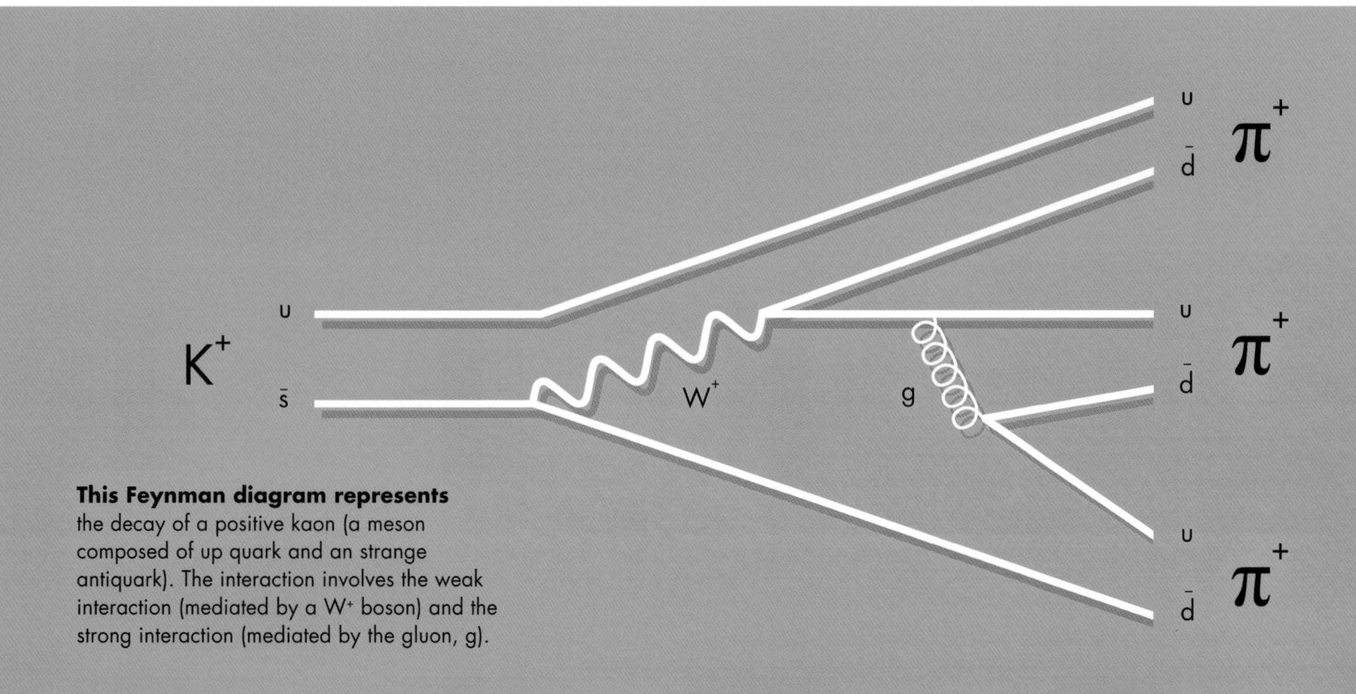

This Feynman diagram represents the decay of a positive kaon (a meson composed of up quark and an strange antiquark). The interaction involves the weak interaction (mediated by a W^+ boson) and the strong interaction (mediated by the gluon, g).

QUANTUM FIELD THEORY

The first theories to address the quantization of the electromagnetic field were made in the late 1920s. They were successful in describing and predicting the interaction between charged particles and photons mathematically. In QED, the photon is a disturbance, or "excitation," that travels through the electromagnetic field, and physicists extended this notion to other interactions beyond electromagnetism. The interactions of particles affected by the strong nuclear force, for example, are described by quantum chromodynamics (QCD). The name derives from the fact that the quarks affected by the powerful nuclear force have a property similar to electric charge, called "color," which determines how they interact. The force-carrying particles in the strong nuclear force are gluons—and in QCD, they exist as quantized excitations of a gluon field, just as photons are quantized excitations of the electromagnetic field. The weak interaction is described by a similar theory called quantum flavor dynamics (QFD—although the weak and electromagnetic interactions have been unified into a single field theory, the electroweak theory).

Quantum field theory (QFT) goes beyond just treating force-carrying particles (photon, gluon, and the W and Z) as excitations of fields. It also treats the particles of matter in the same way. In other words, the electron is actually a quantized excitation of an "electron field" that interacts directly with the electromagnetic field. This makes sense, because according to quantum theory and experiments, the electron behaves as a wave, too. So, according to QFT, there are twelve matter fields: one for each of the six leptons and one for each of the six quarks (each antiparticle is a manifestation of the same field as its particle partner). There are also the four force-carrying fields: one for each of the force-carrying particles (the gluon, the photon, and the W and Z bosons). Then there is the Higgs field—and there may be another, which describes the hypothetical graviton (proposed by some as the carrier of the gravitational interaction). These fields interact with each other, according to well worked-out mathematical rules. These describe not only the structure of matter we experience every day, but also all the many types of creation and annihilation events observed in particle accelerators.

Particle accelerators are a perfect testing ground for QFT. They supply energy that disturbs the fields; disrupt a field with enough energy, and you can create a particle. That was how the Higgs boson—among many other particles—was discovered. Similarly, when an electron loses energy, the electron field passes that energy to the electromagnetic field, and that is how a photon is made. The photons that carry the electromagnetic force (and the other force-carrying gauge bosons) can still affect matter particles without even coming into existence. In fact, most of the time they remain as potential, or "virtual," particles, coming into and out of existence and staying around just long enough for their influence to be felt by the matter particles that they affect. Virtual particles can do this because of a facet of quantum theory called the uncertainty principle.

One frame of a computer- simulated animation shows the fluctuations of quark and gluon fields in empty space. The animation was produced using a supercomputer at the Centre for the Subatomic Structure of Matter and Department of Physics, University of Adelaide, Australia.

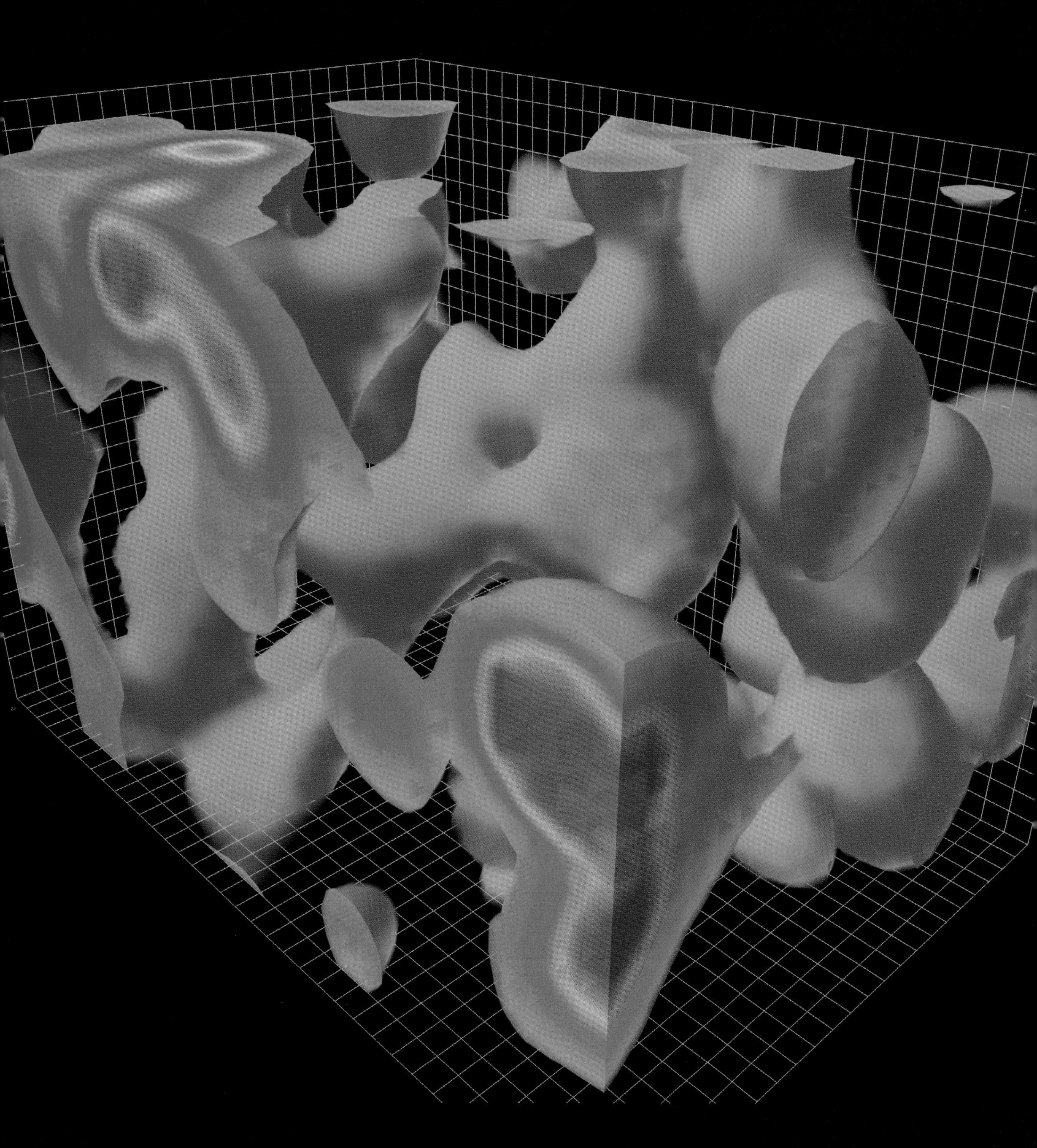

THE UNCERTAINTY PRINCIPLE

In chapter two, we explored two of the pillars of quantum physics—namely, the quantization of energy (and other properties), and wave-particle duality. There is a third: the uncertainty principle. According to the uncertainty principle, it is impossible to know both the exact position and momentum of a particle with complete accuracy (see box below). You can know the position well but have no idea of the momentum, and vice versa, or you can know both within a certain combined mutual limit. This restriction has nothing to do with the accuracy of any measuring apparatus; it is a fundamental law of nature.

The uncertainty relation between position and momentum makes these two properties of a particle "complementary variables." There are other sets of complementary variables, including the energy involved in a process and the duration of that process. It is this fact that makes it entirely possible for a particle to come into existence from nothing, borrowing energy and returning it within a time specified by the uncertainty principle. And because it is possible, it happens. All the matter and force fields that permeate space are constantly seething with virtual particles, their fleeting existence creating a chaotic backdrop to the stage on which the "real" events of the universe take place. The uncertainty principle is also the reason why particles have energy even at absolute zero (see page 132) —and the reason why there is no such thing as truly empty space.

UNIFYING THE FIELDS

Electricity and magnetism were once considered separate forces, but they are now subsumed into a single theory

QUANTUM UNCERTAINTY

The uncertainty principle is a consequence of wave-particle duality and is best explained by considering how to describe a particle in terms of waves. A photon, for example, can be pictured as a wave packet, localized in space. A wave packet can be described mathematically by combining, or superimposing, many waves of different wavelengths. The more different waves you add, the more localized the photon becomes. But each wavelength corresponds to a different momentum, so a particularly localized particle has great uncertainty in its momentum. A photon with a well-defined momentum would have to be represented by a single wave, with a single wavelength, but that wave would not be at all localized in space, stretching out to infinity in both directions.

PRECISELY DEFINED MOMENTUM

The wavelength of a wave function determines a particle's momentum. A particle with a perfectly-defined momentum has a wave function with just a single wavelength. This means it is not at all well-defined in space. In other words, its position cannot be known.

of electromagnetism—and further into a unification with the weak interaction. Theoretical physicists hope to come up with a single theory that can explain all the various quantum fields and show how they are—or once were—part of a single field. In other words, it is possible that all the fields are different versions or manifestations of a single, unified field. In particular, the interactions between particles/fields seem to become unified at particularly high energies, such as those present in the first second of the universe's existence. As the universe cooled, the interactions split into the four seemingly separate interactions that we observe today.

Whether there is a single, unifying quantum field or not, it seems that the only reality is fields. Particles are artifacts created by the excitation of those fields, but they exist within the fields, not separately from them.

American physicist Freeman Dyson eloquently described the view of reality suggested by quantum field theory as long ago as 1953:

"The picture of the world that we have finally reached is the following. Some ten or twenty different quantum fields exist. Each fills the whole of space and has its own particular properties. There is nothing else except these fields: the whole of the material universe is built of them. Between various pairs of fields there are various kinds of interaction. Each field manifests itself as a type of elementary particle. The number of particles of a given type is not fixed, for particles are constantly being created or annihilated or transmuted into one another."

WELL-DEFINED POSITION

A wave function that defines a particle's position precisely, as a wave packet, requires a superposition of many waves of different wavelengths. This means the momentum will be ill-defined.

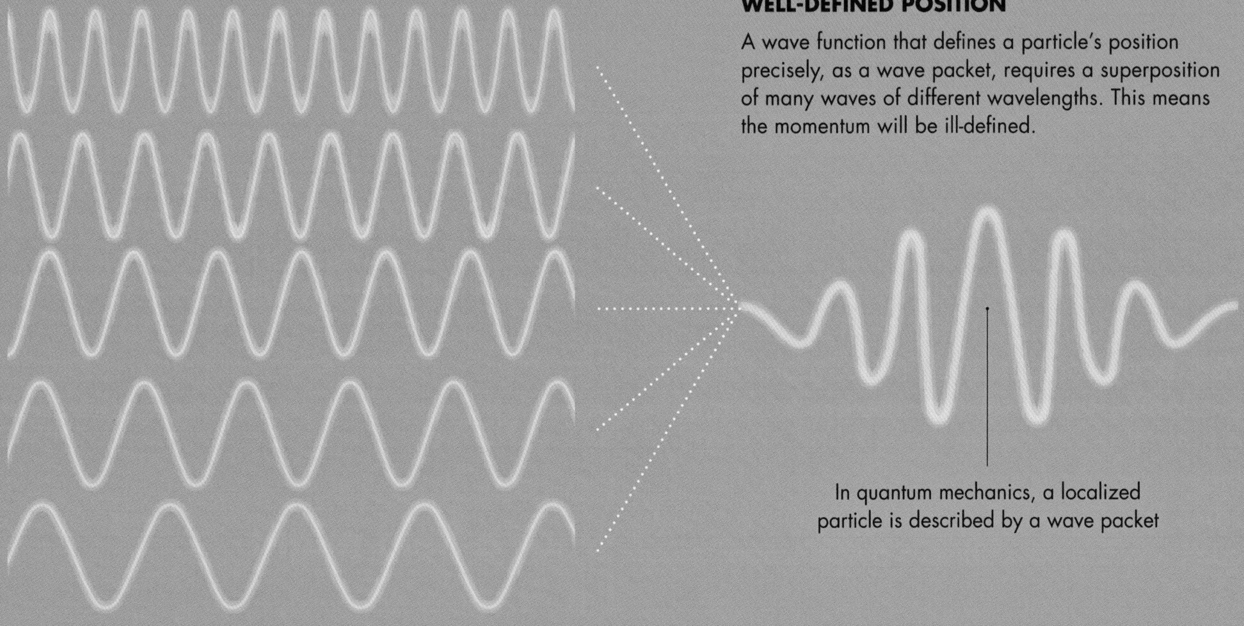

In quantum mechanics, a localized particle is described by a wave packet

A NEW PLENUM

The history of human ideas about what matter is made of has been increasingly dominated by atomism since the scientific revolution in the seventeenth and eighteenth centuries. And yet, in the last sixty years, it has become clear that everything we think of as a solid particle is just an excitation in a field that crosses all of space. The search for the ultimate building blocks of matter has taken us to the conclusion that there is no such thing.

LOOKING BACK

Atoms exist—but they are not the atoms envisaged by Democritus (see page 13). He imagined matter as made of tiny, indivisible balls moving around in empty space. And yet, there is no such thing as empty space; it is filled with fields and a seething, bubbling morass of virtual particles. The modern sense of the word "atom" refers to a composite particle in which electrons are held around a nucleus that is made, ultimately, of triplets of quarks. But even these particles—the electrons and the quarks—are not Democritus's atoms; neither electrons nor quarks are solid balls. Every last one of the ten billion billion billion or so electrons in your body is an excitation of the electron field that fills all of space. Your body contains even more quarks than electrons—and they, too, are excitations of a field. If the quantum fields fill all of space, then everyone's electrons are part of the same field. We are all part of reality's complex, forever-shifting wave function.

Democritus's idea was at odds with his contemporary, the Greek philosopher Parmenides of Elea (see page 12), who was adamant that there is no such thing as empty space, suggesting instead that every last corner of the universe is filled with a substance he called plenum. Nearly two and a half thousand years later, it seems that he was right. It may be hard to accept this picture of reality—comprising simply these ethereal fields existing everywhere—when the world seems to be so solid and weighty. But the solidity of a table and the weight of a bowling ball are both simply caused by interactions between the various fields. In principle, everything can be explained by these interactions.

Of course, no one knows what a field might be "made of." Perhaps, because we can only describe them mathematically, reality is just made of numbers—as Pythagoras and his followers suggested more than two thousand years ago. Or perhaps they are made of some kind of fluidlike substance. After all, if particles travel through fields as quantized waves, then what is doing the waving? If a field really is a substance, then what might that substance be made of? Some kind of atoms, perhaps …?

Of course, no one knows what a field might be "made of."

GLOSSARY

ABSOLUTE ZERO
The lowest possible temperature—at which the kinetic energy of the particles of matter is at a minimum. The value of absolute zero is 0 on the Kelvin temperature scale, equal to –459.67° on the Fahrenheit scale and -273.15° on the Celsius scale.

ALPHA DECAY
The process by which an unstable nucleus attains a lower-energy, more stable state. The nucleus emits an alpha particle, which consists of two protons and two neutrons bound together.

ATOMIC FORCE MICROSCOPY
A technique that produces faithful images of atoms by scanning a surface with an extremely sharp tip and sensing the force between tip and surface.

ATOMIC MASS NUMBER
Often just "mass number," the total number of protons and neutrons in the nucleus. Compare atomic number and atomic weight.

ATOMIC NUMBER
The number of protons in the nucleus of an atom. All the atoms of a particular element have the same atomic number.

ATOMIC WEIGHT
Also called relative atomic mass, the average mass of an atom of a particular element measured in atomic mass units (see dalton). Different isotopes have different atomic masses—this is why the atomic weight is an average.

BETA DECAY
The process by which an unstable nucleus attains a lower-energy, more stable state. Inside the nucleus, a neutron becomes a proton and an electron, the latter being ejected from the nucleus as a beta particle.

BOSE-EINSTEIN CONDENSATE
A state of matter in which bosonic atoms (atoms that can share identical energy quantum states) are cooled to near absolute zero, so that their wave functions merge, and they act as one particle.

BOSON
A particle or group of bound particles with integer spin (0, 1, 2 etc.). Unlike fermions, bosons can be in the same quantum state as other particles of the same kind. Examples are photons, helium-4 atoms and the Higgs boson, also known as the God particle. Compare fermion.

COMPOUND
A substance made of two or more different elements, in which the atoms of the elements are in a definite ratio. An example is water (hydrogen and oxygen, 2:1). The atoms of a compound are bonded, with ionic or covalent bonds.

COVALENT BOND
A bond between atoms in which electrons are shared, in molecular orbitals. A molecule is a group of two or more atoms bonded together covalently.

DALTON (DA)
Also called unified atomic mass units. A measure of atomic and molecular mass. One dalton is equal to one-twelfth the mass of a carbon-12 atom.

DOUBLE SLIT EXPERIMENT
An experiment originally conceived in 1801 to investigate and demonstrate the wave nature of light, but important in modern physics, where it highlights the wave nature of subatomic particles, such as electrons.

ELECTROMAGNETISM
The force between particles that carry electric charge. Along with the strong interaction, the weak interaction and gravity, one of the four fundamental interactions.

ELEMENT
A substance made of one type of atom, defined by the number of protons in the nuclei of those atoms. Examples include hydrogen, oxygen, and carbon.

ELECTRON
A fundamental particle that carries a negative electric charge, found in all atoms.

FERMION
A particle or group of bound particles with half-integer spin (–1/2, 1/2, 3/2 etc.). Unlike bosons, fermions cannot share the same quantum state as other identical particles. Examples are protons, neutrons, electrons and helium-3 atoms. Compare boson.

FIELD EMISSION MICROSCOPY
A technique in which electrons emitted from the surface of a sharp metal point produce a hugely magnified image of the atomic structure at the surface of the tip.

FIELD ION MICROSCOPY
A technique similar to field emission microscopy, in which atoms of a rarefied gas adhere to a sharp metal tip, and are then ionized and expelled from the tip, producing a hugely magnified image of the atoms at the surface of the tip.

GAMMA RAY
A form of electromagnetic radiation with very high frequency (so with very high-energy photons), typically produced during nuclear reactions and radioactive decay.

GLUON
A fundamental particle that carries the strong interaction. Gluons hold together quarks to form protons and neutrons.

GOD PARTICLE
Nickname for the Higgs boson, a particle associated with the Higgs field, a quantum field that is responsible for giving fundamental particles their mass.

HADRON
Any composite particle, made of two or more quarks, that takes part in the strong interaction. The proton and the neutron are hadrons.

HALF-LIFE
The time it takes for half of the atoms of a radioactive isotope to decay.

IONIC BOND
A bond between ions—atoms that have gained or lost electrons. Positive and negative ions held together by ionic bonds typically form crystals.

ISOTOPES
Two or more configurations of the atoms of a particular element, with the same number of protons but differing numbers of neutrons. All elements have at least two isotopes.

LASER
A source of coherent light—light that is of a precise and well-defined wavelength and whose waves are all in phase (in step).

LEPTON
Any particle of matter that, unlike the hadrons, does not take part in the strong interaction. Electrons are leptons.

MAGNETIC RESONANCE IMAGING (MRI)
A medical imaging technique that uses strong magnetic fields and radio waves to interact with the magnetic fields of nuclei.

MAGNETISM
A phenomenon involving particles with spin. In many systems of particles, the spins of the individual particles cancel out, and these systems do not exhibit magnetism.

MASS SPECTROMETRY
An analytical technique that separates mixtures of different molecules by ionising them, accelerating them to high speed, and then bending them with magnetic fields, the amount of deflection determined by their masses.

MOLE
An amount of an element or compound consisting of a number of particles equal to Avogadro's number (600,000 billion billion). One mole of an element has a mass equal to the atomic weight in grams.

MOLECULAR ORBITAL
An orbital formed by the overlap of the orbitals of an atom. As with atomic orbitals, a molecular orbital can hold up to two electrons. Molecular orbitals are the basis of covalent bonds.

NEUTRON
A particle, composed of three quarks, found in all atomic nuclei (except hydrogen-1). Although the quarks carry electric charge, the neutron carries no charge overall.

NUCLEAR FISSION
A nuclear reaction, used in nuclear power stations and weapons, in which large atomic nuclei break, or fission, into two parts, releasing energy. It can happen spontaneously, but can also be made to happen, in the presence of free-moving neutrons.

NUCLEAR FUSION
A nuclear reaction, occurring at the centre of stars and used in thermonuclear weapons, in which small atoms join, or fuse, releasing energy.

NUCLEAR MAGNETISM
The magnetism of nuclei in which the spins of the protons and neutrons do not cancel out. Certain nuclei, including those of hydrogen-1 atoms, have non-zero spin, and therefore act as tiny magnets.

ORBITAL
A region of space, defined by electrons' wave functions, where electrons can be found. Each orbital can hold up to two electrons, with opposing spins.

PARTICLE ACCELERATOR
An apparatus in which subatomic particles or ions are made to travel at high speeds and then collide with each other, creating showers of other particles that allow experimenters to test theories about the subatomic interactions.

PERIODIC TABLE
A chart containing all the elements, arranged in order of increasing atomic number, in rows (periods), arranged so that elements with similar properties—the result of similar electron configurations—are in the same columns (groups).

PHOTOELECTRIC EFFECT
The phenomenon in which illuminating a metal surface with light or other electromagnetic radiation causes electrons to be emitted from the surface. Only radiation with high enough frequency (with photons of high enough energy) can eject electrons.

PHOTON
A fundamental particle. Light and other electromagnetic radiation is a stream of photons, while virtual photons are responsible for carrying the electromagnetic force between charged particles.

PROTON
A particle, composed of three quarks, found in all atomic nuclei. The quarks carry electric charge, and the proton carries an overall positive charge.

QUANTUM THEORY
(Or quantum mechanics or quantum physics.) The scientific theory that describes the behavior of fields, particles and their interactions at the atomic and subatomic scale.

QUANTUM CHROMODYNAMICS
A quantum field theory that describes and predicts the behaviour of particles affected by the strong interaction, notably quarks and gluons.

GLOSSARY

QUANTUM ELECTRODYNAMICS
The quantum field theory that describes and predicts the interactions of photons and particles with electric charge.

QUANTUM FIELD THEORY
The framework by which particles are understood as manifestations of quantized fields that permeate all of space. Each kind of particle in the Standard Model has its own field.

QUARK
A fundamental particle that takes part in the strong interaction. Protons and neutrons are made of quarks (and gluons).

RADIOACTIVITY
(Or radioactive decay.) The process by which unstable atomic nuclei attain lower-energy, often more stable, states. See alpha decay, beta decay, gamma ray.

RADIOCARBON DATING
A technique for estimating the age of once-living matter, by measuring how much of the isotope carbon-14 is present in it. Carbon-14 is taken in at a steady rate by living things, but decays, with a half-life of about 5,700 years.

RELATIVITY
Any theory in physics that acknowledges that matter and energy must behave according to the same fundamental laws wherever they are, whatever speed they are moving at, and however strong is the gravitational field.

SCANNING ELECTRON MICROSCOPY
A technique that produces a magnified image of an object by bouncing electrons, rather than light, off its surface.

SCANNING PROBE MICROSCOPY
Any technique that can be used to produce accurate images of atoms by scanning up and down and sensing the atomic-scale bumps in a surface.

SCANNING TRANSMISSION ELECTRON MICROSCOPY
A technique that produces faithful atomic-scale images by scanning an electron beam up and down across a surface, and collecting the electrons that pass through.

SCANNING TUNNELING MICROSCOPY
A technique that produces faithful atomic-scale images by scanning a probe with a very sharp tip across a surface and measuring the tiny current of electrons that "tunnel" across the gap between tip and surface.

SEMICONDUCTOR
A material that conducts electricity much better than an insulator, but not as well as a metal under normal circumstances. When energized by light or heat, or when doped appropriately, the conductivity can increase.

SPIN
A property of subatomic particles that behaves as if those particles are rotating. Spin gives charged particles magnetism, although paired, opposing spins cancel out. Atoms with unpaired electrons have residual spin, and account for the magnetic properties of certain substances.

STANDARD MODEL
The current best theory that explains the fundamental interactions between subatomic particles.

STRONG INTERACTION
(Or strong force.) The interaction between quarks, carried by virtual gluons. Hadrons (particles composed of quarks) are subject to the strong interaction.

TRIPLE ALPHA PROCESS
A nuclear reaction inside stars in which the would-be carbon-12 nuclei form from three alpha particles.

UNCERTAINTY PRINCIPLE
A major feature of quantum theory, which recognizes fundamental limits on how accurately pairs of quantities can be known—in particular, momentum and position and energy and time.

VIRTUAL PARTICLE
A particle whose fleeting existence is allowed by the uncertainty principle: the particles can exist, from borrowed energy, if the energy is "paid back" within a certain time.

WAVE FUNCTION
A mathematical description of the quantum state of a particle or collection of particles. The value of the wave function at any time and place is related to the probability of a particle being in a particular state at those times and places.

WAVE-PARTICLE DUALITY
The phenomenon in which objects once thought of as particles have wavelike behaviors, and vice versa.

WEAK INTERACTION
(Or weak force.) The interaction between certain subatomic particles, involved in radioactive decay and nuclear reactions.

FURTHER READING

30-Second Quantum Theory
Brian Clegg
(Icon Books, 2014)

Offers a more in-depth treatment of quantum physics without the in-depth mathematics.

Above and Below
Modern Physics for Everyone
Jack Challoner
(Explaining Science Publishing, 2017)

A wide-ranging tour of modern physics. It does not have the depth of *The Atom*, but it covers more ground, including relativity and cosmology.

The Cambridge Guide to the Material World
Rodney Cotterill
(Cambridge University Press, 1989)

This is an old book, and there are parts of it that are a little out of date—but it is a wonderful, intimate, comprehensive look at matter at the atomic scale. It is out of print, but secondhand copies are available online.

The Cell
A Visual Tour of the Building Block of Life
Jack Challoner
(Chicago University Press, 2016)

A sister publication to *The Atom*. Like this book, *The Cell* is a comprehensive but accessible tour of its subject.

QED
The Strange Theory of Light and Matter
Richard Feynman
(Penguin, 1990)

Adapted from a series of lectures on quantum electrodynamics, a theory of physics of which Feynman was a pioneer, this book explains clearly how light and electrons interact. It is the definitive introduction to QED.

Seven Brief Lessons on Physics
Carlo Rovelli
(Penguin, 2016)

This beautifully written book makes a range of complex subjects accessible, in seven easy "lessons." The author is a working theoretical physicist.

ONLINE RESOURCES

QUANTUM PHYSICS I

A free, open, online course presented by the Massachusetts Institute of Technology. A series of in-depth video lectures that require some prior understanding of algebra.

https://ocw.mit.edu/courses/physics/8-04-quantum-physics-i-spring-2013/

or

https://goo.gl/LFB4MW

QUANTUM PHYSICS
by The Khan Academy

A free, online course that explains the principles of quantum physics clearly in a series of videos. Courses on many other areas of science are also available. Highly recommended.
https://www.khanacademy.org/science/physics/quantum-physics

THE DISCOVERY OF THE HIGGS BOSON

A free, open online course by the University of Edinburgh, Scotland, 2018. Only high-school physics understanding is required for this course, which takes the viewer through the steps that led to the discovery of the so-called "God particle."

https://www.class-central.com/course/futurelearn-the-discovery-of-the-higgs-boson-1259
or https://goo.gl/srsU6F

INDEX

F

G

H

I

J

K

L

INDEX

M

N

O

P

Q

R

S

T

U

V

W

X

Y

Z

ACKNOWLEDGMENTS

AUTHOR'S ACKNOWLEDGMENTS

I would like to thank the great team at Ivy Press that put this book together so beautifully—especially the senior editor, Stephanie Evans, designer, Wayne Blades, and copy editor, Catherine Bradley. You all did a great job. Also, Professor Craig Butts at Bristol University, UK, for discussions we had about molecular orbitals.

I produced the molecular and nuclear illustrations using the following free, open source software: Avogadro A molecule editor and visualizer. Available for Windows, Linux and MacOS. https://avogadro.cc. QuteMol High-quality molecular visualization software. Available for Windows and MacOS.

PICTURE CREDITS

The publisher would like to thank the following individuals and organizations for their kind permission to reproduce the images in this book. Every effort has been made to acknowledge the pictures; however, we apologize if there are any unintentional omissions.

Alamy/Jordan Remar: 91; Kropp: 146; Mint Images Limited: 47T; Phil Degginger: 148T; Pix: 111; Science History Images: 26T, 44.

British Library: 14.

Courtesy Francesca Calegari. From 'Ultrafast electron dynamics in phenylalanine initiated by attosecond pulses', F. Calegari et al., *Science* 346, 2014. Reprinted with permission from AAAS.

CERN: 164, 167T, 169T, 171T, 171B, 172, 173T, 174.

Jack Challoner: 57, 59, 63 (nucleus), 70, 71, 73B, 74, 75, 76, 89, 92, 96, 97, 98, 100, 102, 103, 105T, 131, 151, 161.

European Southern Observatory: 104.

Flickr/IPAS/Professor Andre Luiten, adelaide.edu.au, CC-BY-SA: 132; James St Jon, CC-BY: 159; Mdxdt, CC-BY-SA: 118C.

Julie Gagnon, http://www.umop.net/spctelem.htm © 2007, 2013, CC-BY-SA: 68.

Getty Images/Bettmann: 15L; Gallo Images: 60; Guillaume Souvant/AFP: 160; *National Geographic*: 88; Oxford Science Archive/Heritage Images: 19B; Science & Society Picture Library: 21, 26B, 30; Science Photo Library: 15R.

Courtesy Iain Godfrey, SuperSTEM Laboratory, University of Manchester: 121BR.

Viktor Hanacek, picjumbo.com: 12.

Based on an original illustration by Johan Jarnestad/The Royal Swedish Academy of Sciences: 114.

Derek Leinweber, CSSM, University of Adelaide: 178.

Jianwei Miao, University of California, Los Angeles: 112.

NASA: 39BL, 39 (background), 45, 94, 105B, 135T, 157.

National Archives and Records Administration: 161 (background).

NIST: National Institute of Standards and Technology: 124B, 125BL, 129, 135B, 141L, 141R.

Courtesy Quentin Ramasse/Dr. Demie Kepaptsoglou, Prof. Quentin Ramasse, SuperSTEM. Sample from Dr. Vlado Lazarov (University of York) and Sara Majetich (Carnegie Mellon University): 121BL; Dr. Demie Kepaptsoglou, Prof. Quentin Ramasse, SuperSTEM. Samples from Prof. Ursel Bangert, University of Limerick: 121TL; Prof. Quentin Ramasse, SuperSTEM. Sample: Dr. Sigurd Wenner & Prof. Randi Holmestad, NTNU Norway: 121TR.

Science Photo Library: 80, 82C, 82B, 83CL, 83B, 84C, 84BL, 84BR, 85C, 108R; Alfred Pasieka: 83CR; AMMRF/University of Sydney: 118B, 136T, 136B; Brookhaven National Laboratory: 173B; C. Powell, P. Fowler & D. Perkins: 166; Centre Jean Perrin/IBM: 155T; CERN: 170; Charles D. Winters: 90; Don W. Fawcett: 116; Dr. A. Yazdani & Dr. D.J. Hornbaker: 124T; Dr. Mitsuo Ohtsuki: 120; Dr. Kenneth Wheeler: 106; EFDA-JET: 163T, 163B; Emilio Segre Visual Archives/American Institute of Physics: 32; Eye of Science: 125BR, 126R; Gary Cook/Visuals Unlimited: 66L; GIPhotoStock: 27T, 40; Goronwy Tudor Jones/University of Birmingham: 169BL, 169BR; IBM Research: 86, 125T, 127B, 128, 130T, 130B; James King-Holmes: 158; Ken Lucas/Visuals Unlimited: 66BL; Kenneth Eward/Biografx: 2; Martin Land: 66BR; Martyn F. Chillmaid: 66T; NASA's Goddard Space Flight Center/CI Lab: 77; Natural History Museum, London: 66R; NYPL/Science Source: 115R; Omikron: 168; Pascal Goetgheluck: 66BC; Phil Degginger: 27B; Philippe Plailly: 123T, 123B; Prof. D. Skobeltzyn: 167B; Royal Institution of Great Britain: 10; Ted Kinsman: 49B; Victor Shahin, Prof. Dr. H. Oberleithner, University Hospital of Muenster: 127T; Voisin/Phanie: 155B.

Shutterstock/Africa Studio: 39BR; Agrofruti: 108L; Albert Russ: 115L; Alexander Softog: 138; Bjoern Wylezich: 142B; Crafter: 95TL; Dabarti CGI: 150; Gasich Tatiana: 64; Golubovy: 144T; Gopixa: 154; Gustavo Miguel Fernandes: 95BL; HikoPhotography: 95TR; Humdan: 78; Irin-K: 39C; Kai Beercrafter: 20; Kichigin: 95C; L. Nagy: 93; Mikhail Varentsov: 46; Natali art collections: 183; Noor Haswan Noor Azman: 95BR; Pavelis: 109; PNPImages: 101; Speedkingz: 152T; Ugis Riba: 144B.

Image courtesy of Aneta Stodolna. Reprinted with permission from: A. S. Stodolna et al., 'Hydrogen Atoms under Magnification: Direct Observation of the Nodal Structure of Stark States', in *Physical Review Letters*, 110 (21), 213001, May 2013. Copyright 2013 by the American Physical Society.

Wellcome Collection: 18, 19T, 21 (inset), 25, 177T.

Wikimedia Commons/Bdushaw, CC-BY: 62B; Tatsuo Iwata, CC-BY-SA: 117.